国家级实验教学示范中心联席会计算机学科规划教材
教育部高等学校计算机类专业教学指导委员会推荐教材
面向"工程教育认证"计算机系列课程规划教材

WPF编程基础

◎ 刘晋钢 熊风光 况立群

清华大学出版社
北京

内 容 简 介

本书从 WPF 技术的设计原则出发,介绍 WPF 中 XAML 的语法结构、布局方式、常用控件、数据驱动 UI 的理念、路由事件、图形基础、动画与媒体、动作原则、资源与样式及 MVVM 设计模式,并通过大量的案例向读者展示 WPF 的设计思想。案例组织采用分层递进叠加方式,让程序从小变大,由易到难,能够使读者迅速地熟悉编写程序的思想路径,体会到编写程序的乐趣。每章配有习题,以启发读者深入地学习 WPF 技术。

本书既可作为高等学校计算机专业 UI 设计、软件开发、人机交互技术等课程的教材,也可作为计算机从业人员的参考书。

本书封面贴有清华大学出版社防伪标签,无标签者不得销售。
版权所有,侵权必究。举报: 010-62782989,beiqinquan@tup.tsinghua.edu.cn。

图书在版编目(CIP)数据

WPF 编程基础/刘晋钢等编著. —北京: 清华大学出版社,2018(2024.12重印)
(面向"工程教育认证"计算机系列课程规划教材)
ISBN 978-7-302-48281-9

Ⅰ. ①W… Ⅱ. ①刘… Ⅲ. ①Windows 操作系统-程序设计-高等学校-教材 Ⅳ. ①TP316.7

中国版本图书馆 CIP 数据核字(2017)第 208267 号

责任编辑: 付弘宇　王冰飞
封面设计: 刘　键
责任校对: 李建庄
责任印制: 丛怀宇

出版发行: 清华大学出版社
　　　网　　址: https://www.tup.com.cn, https://www.wqxuetang.com
　　　地　　址: 北京清华大学学研大厦 A 座　　　邮　编: 100084
　　　社 总 机: 010-83470000　　　邮　购: 010-62786544
　　　投稿与读者服务: 010-62776969, c-service@tup.tsinghua.edu.cn
　　　质量反馈: 010-62772015, zhiliang@tup.tsinghua.edu.cn
　　　课件下载: https://www.tup.com.cn, 010-83470236
印 装 者: 北京鑫海金澳胶印有限公司
经　　销: 全国新华书店
开　　本: 185mm×260mm　　　印　张: 15.75　　　字　数: 383 千字
版　　次: 2018 年 3 月第 1 版　　　印　次: 2024 年 12 月第 8 次印刷
印　　数: 5101~5600
定　　价: 49.00 元

产品编号: 074796-02

前 言

WPF(Windows Presentation Foundation)是专门用来编写程序表示层的技术和工具。它是微软新一代图形展示系统,是用户界面技术进步的重要标志。使用 WPF 编写的程序比之前的 WinForm 程序更加简洁清晰。WPF 技术适用于微软平台下的桌面系统、浏览器、Windows Phone 的开发。因为微软程序的开发理念都一样,仅在类库方面有一些差别。

本书详细介绍 WPF 中 XAML 的语法结构、布局方式、常用控件、数据驱动 UI 的理念、路由事件、图形基础、动画与媒体、动作原则、资源与样式及 MVVM 设计模式,并通过大量的案例向读者展示 WPF 的设计思想。案例组织采用分层递进叠加方式,让程序从小变大,由易到难,能够使读者迅速地熟悉编写程序的思想路径,体会到编写程序的乐趣。全书共 12 章,前 4 章是有关 WPF 基础的编程内容和界面 UI 设计,从第 5 章开始是 WPF 的高级进阶。各章内容概述如下。

第 1 章介绍 WPF 的编程机制。采用逐层深入的 Button 案例,讲解 WPF 平台特性。通过对 WPF 的运行机制及类层次结构讲解,从而认识 WPF 的体系结构。

第 2 章介绍 XAML 可扩展的应用程序声明式语言的树形结构、复杂属性、附加属性、xmlns 指令和名称空间中的标记扩展等。

第 3 章详细介绍 WPF 布局原则及各布局面板的适用场合。重点说明 Grid 从结构中分离布局、尺寸模型、共享尺寸组、跨越行列等特征,并演示了 Grid 的多种用法。

第 4 章介绍 WPF 控件内容模型和模板的新概念。重点说明元素合成、富内容和简单的编程模型的控件原则。在此基础上,学习 WPF 的内置控件。

第 5 章介绍数据驱动模型、数据绑定原理及数据绑定的用法。

第 6 章从 Windows 操作系统的消息机制出发,介绍事件模型。在 WPF 中引入路由事件机制,可采用冒泡、隧道、直接 3 种策略。

第 7 章从常用的几何图形元素出发,介绍绘制图画、2D 形状及属性,让读者进一步认识 WPF 3D 三维空间坐标系、模型、材质、光源、照相机和变换。

第 8 章介绍动画工作原理。WPF 动画根据计算机的性能和当前进程的繁忙程度,尽可能地增大帧率,比传统动画流畅,实现方式简捷。同时还介绍 WPF 中动画的常用类型、集成方式和对音频与视频文件的播放方式。

第 9 章介绍动作使用原则、命令系统及触发器使用方式。

第 10 章介绍资源的定义、资源类型、适用范围、资源的静态与动态引用方式。特别强调资源字典要遵循"先创建后使用"的规则。

第 11 章重点介绍样式的构成、如何使用样式及模板，并演示 WPF 实现换肤的操作步骤。

第 12 章介绍软件设计模式的起源、概念和原则。对比 MVC、MVP 和 MVVM 3 种设计模式的通信方式的差别，重点讲解 MVVM 设计模式的框架、其三大组件内容结构，用基于 MVVM 模式的计算器案例证明 WPF 数据驱动 UI 的设计思想。

本书由刘晋钢主编，其中第 1、4、12 章由刘晋钢编写，第 2、3 章由熊风光编写，第 6、7 章由况立群编写，第 8、9 章由刘晋霞编写，第 10、11 章由张麟华编写，第 5 章由李丽编写。

本书既可作为高等学校计算机专业学生用书，也可作为计算机从业人员的参考书。作为教材，本书既适用于计算机专业 UI 设计、软件开发、人机交互技术等课程，也可供职业技术学院计算机专业学生使用。通过本书的学习，读者可以掌握 WPF 的核心技术，提升程序设计能力，为以后的工作和研究打下坚实的基础。

读者可在清华大学出版社网站（www.tup.com.cn）免费下载本书所有案例的源代码、与本书配套的电子课件以及习题参考答案。使用本书时，遇到资源下载问题，请联系责任编辑 fuhy@tup.tsinghua.edu.cn 或联系本书作者 84161924@qq.com。

在此，特别感谢刘子民对本书提供的技术支持和帮助，也感谢本书的责任编辑付弘宇对本书所做的审核工作。由于编者的水平有限，书中难免存在不足之处，恳请广大读者批评指正。

教学建议

根据突出应用的原则，从应用层次要求角度考虑，可把"WPF 编程基础"课程的教学内容分为基础学习内容、高级进阶内容和提升应用研发能力三部分。

基础学习：本书的前 4 章内容是有关 WPF 基础的编程内容和界面 UI 设计。详细介绍 WPF 中 XAML 的语法结构、布局方式和常用控件，涵盖 WPF 的新特性、体系结构、XAML 基础语法知识、XAML 文档的树形结构和常用属性，以及 WPF 布局原则及布局面板、WPF 控件模型、模板及常用控件。学完前 4 章，读者可以做出赏心悦目的用户界面。

高级进阶：本书的第 5 章到第 9 章是 WPF 的高级进阶。详细介绍数据驱动 UI 的理念、图形基础、动画与媒体和动作原则。这部分内容涉及 WPF 的核心技术，将事件驱动模型提升到数据驱动 UI 的理念上来，让 UI 与业务逻辑真正地分离，并使前台的设计师和后台的程序员各司其职。

提升应用：本书的第 10 章到第 12 章是提升学生应用能力部分。详细介绍资源的类型、引用方式、资源字典、样式的构成、使用样式的方法、模板及 MVVM 设计模式。目前学生开发的不少项目多半都废弃，究其原因主要是，资源分配不合理，样式不美观，没有采用好的设计模式。这部分内容针对上述问题编写，以提高学生的应用研发能力。

本书用于教学的建议如表 1 所示。

表 1 教学建议

课程名 （授课对象）	WPF 编程基础（授课对象可以是计算机科学与技术、软件工程等专业理工类四年制本科）48 学时。注：三年制专科可参考此计划适当修改（例如，可延长到 64 学时，难度适当降低）
教学目的和要求	掌握 WPF 核心技术、体系结构、数据驱动 UI 的设计思想及 MVVM 设计模式，为提升程序开发能力奠定基础

续表

必要的先修课程	数据结构、操作系统和面向对象程序设计语言(如C#)		
后续可开设实践课程	WPF项目实训、基于Kinect的体感设计及手势识别项目研发		
章次(学时)	要求学生了解内容	要求学生掌握内容	实践操作项目
第1章 引言(2)	WPF的地位、体系结构及应用前景	WPF的新特性,布局与控件中涉及的容器控件的简单用法	数据集成处理能力章节中的Button导出的案例
第2章 XAML(2)	XAML与HTML的异同点、XAML文档的树形结构	XAML的名称空间及属性、类型转换器的用法	仿类型转换器中案例,重做一个将字符串转换成对象实例
第3章 布局(4)	合成布局模型、布局机制、布局常用属性	布局面板的用法,布局嵌套	生活中常见的布局应用(聊天室、Web等)
第4章 控件(4)	元素合成、富内容和简单的编程模型的控件原则	WPF控件的基本用法、构建控件的思想、用户自定义控件	图标设计、登录用户页面、游戏初始化页面、桌面、主题页面等
第5章 数据(6)	数据模型的发展过程及微软曾用过的数据模型	数据绑定机制、值转换机制、数据绑定模型、数据绑定用法	INotifyPropertyChanged接口调用、数据绑定列表框
第6章 路由事件(6)	消息概念、消息循环、Windows编程原理、附加事件	路由事件工作机制、RoutedEventArgs类、路由策略	实现自定义路由事件,分别采用隧道、冒泡和直接3种策略
第7章 图形基础(4)	常用几何图形、在WPF 3D中的基本概念,包括WPF的坐标系、各种光源和照相机的工作原理	WPF图像特效、绘制图画、控件与形状组合、常用变换	使用MeshGeometry3D定义模型,创建三维物体
第8章 动画与媒体(2)	动画的概念、原理,传统动画与WPF动画异同点	线性插值动画、关键帧动画、路径动画,动画集成	设计简单的动画
第9章 动作(6)	动作原则、命令系统的基本元素及元素间的关系、WPF命令库	使用各种触发器、命令与数据绑定	实现Windows记事本功能
第10章 资源(4)	资源的定义、资源的类型、资源的可用范围、使用资源的意义	资源的静态引用与动态引用方式、创建和使用资源字典	使用资源字典
第11章 样式(4)	样式的作用	定制模板、使用样式的方法	设置主题、锁屏、更换壁纸(皮肤)
第12章 MVVM设计模式(4)	软件设计模式的概念、原则、由来,MVC、MVP和MVVM发展过程	MVVM设计模式的框架及其三大组件内容结构及该模式的优点	完成基于MVVM的简单计算器,并为该计算器的按钮设计统一风格的样式

编者

2018年1月

目 录

第 1 章 引言 ·· 1
  1.1 全新的图形用户系统 ··· 1
  1.2 XAML 编程模型 ·· 2
    1.2.1 HTML ··· 2
    1.2.2 XAML ··· 3
  1.3 WPF 特性 ·· 3
    1.3.1 布局与控件 ·· 5
    1.3.2 数据集成及处理能力 ·· 6
  1.4 WPF 体系结构 ··· 13
    1.4.1 WPF 运行机制 ··· 13
    1.4.2 WPF 类层次结构 ·· 14
    1.4.3 WPF 的可视化树与逻辑树 ·· 15
  1.5 WPF 与 UWP ·· 16
  1.6 小结 ·· 17
  习题与实验 1 ·· 17

第 2 章 XAML ·· 19
  2.1 XAML 文档框架 ·· 19
    2.1.1 XAML 文档结构 ·· 20
    2.1.2 基础语法 ·· 20
  2.2 XAML 中的属性 ·· 21
    2.2.1 简单属性 ·· 21
    2.2.2 复杂属性 ·· 22
    2.2.3 附加属性 ·· 23
    2.2.4 处理特殊字符与空白 ··· 23
  2.3 XAML 名称空间 ·· 24

2.3.1　名称空间的作用 ··· 24
　　　2.3.2　默认名称空间 ··· 25
　　　2.3.3　名称空间中的标记扩展 ·· 25
　2.4　类型转换器 ·· 26
　2.5　导入程序集 ·· 28
　2.6　小结 ··· 29
　习题与实验 2 ·· 29

第 3 章　布局 ·· 31

　3.1　布局原则 ·· 31
　　　3.1.1　合成布局模型 ··· 31
　　　3.1.2　布局机制 ··· 32
　　　3.1.3　布局通用属性 ··· 33
　3.2　布局面板 ·· 33
　　　3.2.1　Canvas ·· 34
　　　3.2.2　DockPanel ··· 35
　　　3.2.3　StackPanel ··· 37
　　　3.2.4　WrapPanel ··· 38
　　　3.2.5　UniformGrid ·· 39
　3.3　Grid ·· 40
　　　3.3.1　从结构中分离布局 ··· 42
　　　3.3.2　尺寸模型 ··· 43
　　　3.3.3　共享尺寸组 ·· 45
　　　3.3.4　跨越行和列 ·· 45
　　　3.3.5　GridSplitter ·· 47
　3.4　小结 ··· 48
　习题与实验 3 ·· 49

第 4 章　控件 ·· 50

　4.1　WPF 控件新理念 ·· 50
　　　4.1.1　内容模型 ··· 50
　　　4.1.2　模板 ··· 53
　4.2　菜单、工具栏和状态栏 ··· 57
　　　4.2.1　Menu ·· 57
　　　4.2.2　ToolBar ··· 58
　　　4.2.3　StatusBar ··· 59
　4.3　容器控件 ·· 60

		4.3.1	Expander ···	60
		4.3.2	GroupBox ··	61
		4.3.3	TabControl ··	62
	4.4	范围控件 ···		62
		4.4.1	Slider ··	63
		4.4.2	ScrollBar ··	64
		4.4.3	ProgressBar ···	64
	4.5	文本编辑器控件 ··		64
		4.5.1	文本模型 ··	65
		4.5.2	PasswordBox ··	65
		4.5.3	TextBox 与 RichTextBox ··	65
		4.5.4	InkCanvas ···	66
	4.6	列表控件 ···		68
		4.6.1	ListBox 和 ComboBox ···	68
		4.6.2	ListView ···	70
		4.6.3	TreeView ··	71
	4.7	构建控件 ···		73
		4.7.1	ToolTip ··	73
		4.7.2	Thumb ···	74
		4.7.3	Border ···	76
		4.7.4	Popup ···	77
		4.7.5	ScrollViewer ···	79
		4.7.6	Viewbox ···	79
	4.8	日期控件 ···		80
		4.8.1	Calendar ···	80
		4.8.2	DatePicker ··	81
	4.9	按钮 ···		82
	4.10	小结 ···		83
	习题与实验 4 ···			83
第 5 章	数据 ···			85
	5.1	数据驱动模型 ··		85
		5.1.1	数据原则 ··	85
		5.1.2	资源 ··	86
	5.2	数据绑定原理 ··		87
		5.2.1	数据绑定机制 ···	87
		5.2.2	数据源与路径 ···	91

         5.2.3　值转换机制 ··· 93
         5.2.4　数据绑定模型 ··· 95
   5.3　数据绑定用法 ·· 97
         5.3.1　控件间的绑定 ··· 97
         5.3.2　控件绑定资源文件值 ··· 98
         5.3.3　属性变更通知接口 ··· 99
         5.3.4　绑定到列表框 ··· 101
   5.4　小结 ·· 104
   习题与实验 5 ··· 104

第 6 章　路由事件 ·· 106
   6.1　消息机制 ·· 106
         6.1.1　消息的运行机制 ·· 106
         6.1.2　事件模型 ·· 109
   6.2　路由事件原理 ··· 110
         6.2.1　路由事件机制 ··· 110
         6.2.2　RoutedEventArgs 类 ··· 115
         6.2.3　路由策略 ·· 115
   6.3　自定义路由事件 ·· 117
   6.4　附加事件 ·· 120
   6.5　小结 ·· 120
   习题与实验 6 ··· 120

第 7 章　图形基础 ·· 122
   7.1　WPF 图形原则 ·· 122
         7.1.1　几何图形与笔刷 ·· 122
         7.1.2　绘制图画 ·· 128
   7.2　2D 图形 ·· 131
         7.2.1　形状 ··· 131
         7.2.2　图像 ··· 134
         7.2.3　WPF 图像特效 ·· 137
   7.3　3D 图形 ·· 139
         7.3.1　WPF 坐标系 ·· 139
         7.3.2　模型 ··· 141
         7.3.3　材质 ··· 145
         7.3.4　光源与照相机 ··· 146
         7.3.5　变换 ··· 147

7.4 小结 …………………………………………………………………………… 148

习题与实验 7 ………………………………………………………………………… 148

第 8 章 动画与媒体 ……………………………………………………………………… 150

8.1 动画基础 ……………………………………………………………………… 150

8.1.1 动画的概念 ……………………………………………………… 150

8.1.2 动画的原理 ……………………………………………………… 150

8.1.3 传统动画与 WPF 动画 ………………………………………… 151

8.2 动画类型 ……………………………………………………………………… 152

8.2.1 线性插值动画 …………………………………………………… 152

8.2.2 关键帧动画 ……………………………………………………… 154

8.2.3 路径动画 ………………………………………………………… 155

8.3 集成动画 ……………………………………………………………………… 158

8.3.1 与控件模板集成 ………………………………………………… 158

8.3.2 与文本类型集成 ………………………………………………… 159

8.4 媒体 …………………………………………………………………………… 160

8.4.1 音频 ……………………………………………………………… 160

8.4.2 视频 ……………………………………………………………… 161

8.5 小结 …………………………………………………………………………… 162

习题与实验 8 ………………………………………………………………………… 162

第 9 章 动作 ……………………………………………………………………………… 164

9.1 动作原则 ……………………………………………………………………… 164

9.1.1 元素合成 ………………………………………………………… 164

9.1.2 松散耦合 ………………………………………………………… 165

9.1.3 声明式动作 ……………………………………………………… 166

9.2 命令系统 ……………………………………………………………………… 166

9.2.1 基本元素及元素之间的关系 …………………………………… 166

9.2.2 ICommand 接口 ………………………………………………… 170

9.2.3 RoutedCommand 类 …………………………………………… 171

9.2.4 RoutedUICommand 类 ………………………………………… 171

9.2.5 WPF 命令库 …………………………………………………… 171

9.2.6 命令与数据绑定 ………………………………………………… 172

9.3 触发器 ………………………………………………………………………… 175

9.3.1 数据触发器 ……………………………………………………… 176

9.3.2 属性触发器 ……………………………………………………… 178

9.3.3 多条件触发器 …………………………………………………… 179

9.4 小结 ……………………………………………………………………… 182
习题与实验 9 …………………………………………………………………… 182

第 10 章 资源 …………………………………………………………………… 183

10.1 资源概述 ………………………………………………………………… 183
    10.1.1 资源的定义 ……………………………………………………… 183
    10.1.2 资源可用范围 …………………………………………………… 184
10.2 资源类型 ………………………………………………………………… 185
    10.2.1 二进制资源 ……………………………………………………… 185
    10.2.2 逻辑资源 ………………………………………………………… 188
10.3 资源引用方式 …………………………………………………………… 188
    10.3.1 静态资源引用 …………………………………………………… 189
    10.3.2 动态资源引用 …………………………………………………… 189
10.4 资源字典 ………………………………………………………………… 190
    10.4.1 创建资源字典 …………………………………………………… 190
    10.4.2 使用资源字典 …………………………………………………… 191
10.5 小结 ……………………………………………………………………… 193
习题与实验 10 ………………………………………………………………… 194

第 11 章 样式 …………………………………………………………………… 195

11.1 样式的构成 ……………………………………………………………… 195
    11.1.1 设置器 …………………………………………………………… 195
    11.1.2 样式触发器 ……………………………………………………… 195
    11.1.3 样式容器 ………………………………………………………… 198
11.2 使用样式的方法 ………………………………………………………… 198
    11.2.1 内联样式 ………………………………………………………… 198
    11.2.2 已命名样式 ……………………………………………………… 199
    11.2.3 元素类型样式 …………………………………………………… 201
    11.2.4 编程控制样式 …………………………………………………… 203
11.3 模板 ……………………………………………………………………… 204
    11.3.1 定制模板 ………………………………………………………… 204
    11.3.2 样式与控件模板 ………………………………………………… 206
    11.3.3 样式与数据模板 ………………………………………………… 207
    11.3.4 列表与项目模板 ………………………………………………… 210
    11.3.5 主题与皮肤 ……………………………………………………… 212
11.4 小结 ……………………………………………………………………… 214
习题与实验 11 ………………………………………………………………… 214

第 12 章　MVVM 设计模式 ·········· 217

12.1　软件设计模式 ·········· 217
12.1.1　设计模式的概念 ·········· 217
12.1.2　设计模式的原则 ·········· 217

12.2　MVVM 设计模式概述 ·········· 218
12.2.1　MVVM 的由来 ·········· 218
12.2.2　MVVM 框架 ·········· 220
12.2.3　MVVM 的优点 ·········· 221

12.3　基于 MVVM 的计算器设计 ·········· 221
12.3.1　Model ·········· 223
12.3.2　ViewModel ·········· 224
12.3.3　View ·········· 228

12.4　基于 MVVM 设计思想 ·········· 235
12.5　小结 ·········· 236
习题与实验 12 ·········· 236

参考文献 ·········· 237

第1章 引言

WPF(Windows Presentation Foundation)是微软新一代图形展示系统,是用户界面技术进步的重要标志。本章向读者呈现 WPF 平台的主要功能,也可视作本书其他章节的缩写版。对于初学者,第 1 章有一定的难度,因此可以从第 2 章学起,借助后续分层递进式案例,回过头来再阅读本章内容,会带给读者一种豁然开朗的感觉。

1.1 全新的图形用户系统

WPF 提供统一的编程模型、语言和框架,还有全新的多媒体交互用户图形界面,如图 1.1 所示。

图 1.1 WPF DEMO

WPF DEMO 集成了类似于 Windows 8 样式按钮的风格,支持触摸以及数据绑定。下面首先回顾 Windows Form 托管对象模型下的"Hello Form"程序,代码如下。

```
using System;
using System.Windows.Forms;
namespace WindowsFormsApplication1
{   static class Program
    {   [STAThread]
        static void Main()
        {   Form form = new Form();
            form.Text = "Hello Form";
            Application.Run(form);
        }
    }
}
```

尽管 WPF 被视为全新的图形用户系统,但它还是保留着与 Windows Form 程序类似的风格,对上述代码做部分修改。

```
using System;
using System.Windows;
namespace WPF
{    static class Program
     {      [STAThread]
        static void Main()
        {   Window form = new Window();
            form.Title = "Hello Form";
            new Application().Run(form);
        }
     }
}
```

在 WPF 环境下,该程序无语法错误,但是编译时报错,错误类型是编译后的.obj 文件中也包含了一个 Main(),程序出现了两个入口,不能运行。众所周知,Main()是 C♯ 程序的入口。在后续的学习中,读者将了解到,WPF 程序默认的启动窗体是 MainWindow()。也许读者会有这样的疑问,编译时生成的 Main()和 MainWindow()是不是有冲突呢?其实 WPF 的入口并不是 MainWindow,而是 App.xaml 中的 App 类。第 12 章将通过案例让读者再深入理解 WPF 的运行机制。

WPF 需运行在.NET Framework 3.0 以上版本,给用户界面、2D/3D 图形、文档和媒体提供统一的描述和操作方法,并且支持 DirectX 9/10 技术。

WPF 是基于矢量(Vector-based)的合成系统,这就意味着它支持旋转、缩放等各种变换,这些变换均可作用于控件,使控件可以正常工作。

1.2　XAML 编程模型

XAML(eXtensible Application Markup Language,可扩展应用程序标记语言)编程模型,类似于 HTML(Hyper Text Markup Language,超级文本标记语言)编程模型,它继承了 Web(Website 的缩写)开发的特性。Web 开发的优点是:为内容创建简单的入口,基于 HTML 编程。下面说明 HTML 与 XAML 两者之间的异同点。

1.2.1　HTML

HTML 文档制作简单,但功能强大,支持多种数据格式,它具有简单灵活、平台无关性、通用性强等多种优点,在文本文件中即可创建 HTML 标签,并将该文件命名为"welcome.html",代码如下。

```
<html>
   <head>
     <title> hello Everyone </title>
   </head>
     <body>
       <p> Welcome to HTML! </p>
     </body>
</html>
```

诚然，上面所有的标签都可以省略，只输入一行文字 Welcome to HTML，运行后，可看到的浏览器模式效果如图 1.2 所示。

1.2.2 XAML

在 WPF 中，使用 XAML 的标记格式，它与 HTML 在语法结构上有许多相似之处，但两者之间最显著的不同特征是：xmlns 指令和标记扩展。

其中 xmlns 指令，需要在标记中关联名称空间，在文本文件中即可创建 XAML 标签，并将该文件命名为"welcome.xaml"，代码如下。

```
<FlowDocument
    xmlns = "http://schemas.microsoft.com/winfx/2006/xaml/presentation">
    <Paragraph>Welcome to XAML</Paragraph>
</FlowDocument>
```

双击该文件，浏览器中的显示效果如图 1.3 所示。

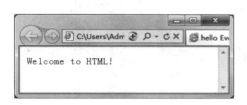

图 1.2　运行 welcome.html 效果

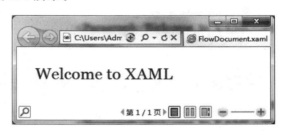

图 1.3　运行 welcome.xaml 效果

由于 WPF 具有 xmlns 指令，需要在标记中关联名称空间，故 Microsoft 的.NET Framework 3.0 就已经具有 XAML 的编程模型，支持 XML(eXtensible Markup Language,可扩展置标语言)语法，对界面进行解释。因此，XAML 可以视为针对 CLR(Common Language Runtime,公共语言运行环境)对象编写。它基于 XML 的脚本语言，根据映射规则将 XML 标签转换为 CLR 类型，并把 XML 属性转换为 CLR 属性和事件。在后续的章节中，展开讲解使用 XAML 和 C♯(在后面章节中常提到的 CS 代码就特指 C♯)创建对象并设置其属性的方法。

XAML 的标记扩展新特性，增加了程序的可读性，标记扩展以 CLR 类型实现，工作方式与 CLR 属性定义一致。标记扩展放在花括号{}中，将在第 2 章中进行介绍，这里不再赘述。

大家知道，在 XML 脚本语言中，只存在元素和属性两个空间，但是，在 XAML 模型中，有对象、属性和事件 3 种描述，接近于 CLR 方式。这样，WPF 应用程序可由 Microsoft Visual Studio 和 Microsoft Expression 这些优秀的可视化设计工具来创建。

对 XAML 的编程模型大致了解后，下面介绍 WPF 的新特性。

1.3　WPF 特性

WPF 开发有很多种方式，包括标记方式、浏览器方式和编码方式。因为 Microsoft Visual Studio 提供可视化设计工具，从本节开始，书中的所有程序都将运行在 Microsoft

Visual Studio 2010 平台下。

在创建第一个 WPF 应用程序之前,先启动 Visual Studio 2010 软件。使用"文件"→"新建"→"项目"菜单命令时,会弹出新建 WPF 应用程序窗口,如图 1.4 所示。在该窗口左侧模板列表中选择 Visual C# 选项;在窗口中间的下拉列表项中选择 .NET Framework 4 选项;并选择其下的"WPF 应用程序"选项。在名称文本框中输入 Button_0,此时 Button_0 就是解决方案名称。

图 1.4　新建 WPF 应用程序窗口

在图 1.4 中单击"确定"按钮后,弹出 WPF 应用程序开发窗口,如图 1.5 所示。

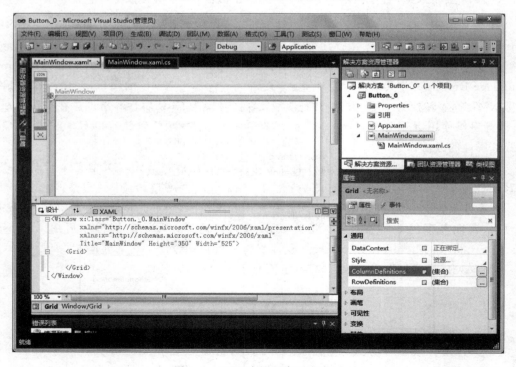

图 1.5　WPF 应用程序开发窗口

在图 1.5 左上部分有 MainWindow.xaml 和 MainWindow.xaml.cs 两个选项卡,前者是用于设计前台 UI 的 XAML 文档;后者是负责后台业务逻辑的 CS 文档。当前状态下,系统默认 XAML 文档为当前的工作界面,所以 MainWindow.xaml 高亮显示,在窗口左下方是其编码界面。在窗口右上方是"解决方案管理器"。

1.3.1 布局与控件

WPF 应用程序的开发是和控件关联在一起的。新建 WPF 应用程序,命名为"Button._0",进入启动页面后,在解决方案管理器下,双击 MainWindow.xaml 文件名,打开该文件。该文件是系统自动生成的。首先看到了 Window 对象,它就是 WPF 的控件。下面从大家最熟悉的控件<Button>开始来介绍 WPF 布局与控件。

在<Grid>与</Grid>之间输入<Button>First button</Button>,代码如下。

```
<Window x:Class = "Button._0.MainWindow"
        xmlns = "http://schemas.microsoft.com/winfx/2006/xaml/presentation"
        xmlns:x = "http://schemas.microsoft.com/winfx/2006/xaml"
        Title = "MainWindow" Height = "350" Width = "525">
    <Grid>
        <Button>First Button</Button>
    </Grid>
</Window>
```

运行这段代码,产生如图 1.6 所示的效果,可以看到按钮充满窗口整个区域,因为使用了 WPF 默认的 Grid 布局。

在 WPF 中所有的控件都有规定好的布局类型。因此,为了在窗口中放置更多的控件,需要使用容器控件。将<Grid>更换成<StackPanel>布局后,再加入一个<Button>,代码如下。

```
<Window x:Class = "Button._0.MainWindow"
        xmlns = "http://schemas.microsoft.com/winfx/2006/xaml/presentation"
        xmlns:x = "http://schemas.microsoft.com/winfx/2006/xaml"
        Title = "MainWindow" Height = "350" Width = "525">
    <StackPanel>
        <Button>First Button</Button>
        <Button>Second Button</Button>
    </StackPanel>
</Window>
```

<StackPanel>按照控件出现的先后顺序,从上往下摆放控件,进行布局,运行效果如图 1.7 所示。

图 1.6 Grid 布局窗口 Button

图 1.7 StackPanel 布局两个 Button

为了深入理解<StackPanel>的布局方式,再加入<TextBox>和<TextBlock>两控件,代码如下。

```xml
<Window x:Class = "Button._0.MainWindow"
        xmlns = "http://schemas.microsoft.com/winfx/2006/xaml/presentation"
        xmlns:x = "http://schemas.microsoft.com/winfx/2006/xaml"
        Title = "MainWindow" Height = "350" Width = "525">
    <StackPanel>
        <Button>First Button</Button>
        <Button>Second Button</Button>
        <TextBox>I am a TextBox</TextBox>
        <TextBlock>I am a TextBlock</TextBlock>
    </StackPanel>
</Window>
```

运行上面的代码,显示页面效果如图1.8所示。

下面更换布局,将<StackPanel>替换成<WrapPanel>,代码如下。

```xml
<Window x:Class = "Button._0.MainWindow"
        xmlns = "http://schemas.microsoft.com/winfx/2006/xaml/presentation"
        xmlns:x = "http://schemas.microsoft.com/winfx/2006/xaml"
        Title = "MainWindow" Height = "350" Width = "525">
    <WrapPanel>
        <Button>First Button</Button>
        <Button>Second Button</Button>
        <TextBox>I am a TextBox</TextBox>
        <TextBlock>I am a TextBlock</TextBlock>
    </WrapPanel>
</Window>
```

<WrapPanel>按照控件出现的先后顺序,从左往右摆放控件,进行布局,运行效果如图1.9所示。

图1.8 StackPanel布局的多个控件排列

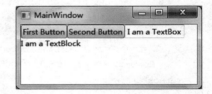

图1.9 WrapPanel布局的多控件排列

通过上述简单案例,读者初步了解了WPF的布局和控件,下面来认识一下WPF的数据处理能力。

1.3.2 数据集成及处理能力

1. 数据绑定机制

WPF实现控件绑定数据源,并提供多种数据绑定的方法。在此,继续使用上述Button示例。

Button 绑定类型有 3 种。第一，绑定类型决定按钮的显示方式。所有的控件都有一个 Resources(资源)属性。Resources 包括 Style(样式)、Templates(模板)与 Resources Dictionary(资源字典)。第二，按钮内容的数据类型是 System.Object 对象，按钮能获取任何数据(包括图片)，并显示内容。WPF 的许多控件核心就是 Content Model(内容模型)机制。第三，数据绑定来实现 Content Model(内容模型)。

在上例的基础上，添加代码来实现数据绑定。案例设计方案是：将按钮的背景设成黄色，当然，也可以给每个按钮的 Background 属性设成黄色，但是如果应用程序中有许多按钮，按钮的风格要保持一致，都设成黄色。最简单的方式就是把颜色定义放到 Resources 属性中，让所有的按钮都关联该属性。

在 Windows 中定义一个 Resources，在控件的 Resources 中声明一个对象，并为这个对象分配一个"x:Key"，代码如下。

```xml
< Window x:Class = "Button._0.MainWindow"
        xmlns = "http://schemas.microsoft.com/winfx/2006/xaml/presentation"
        xmlns:x = "http://schemas.microsoft.com/winfx/2006/xaml"
        Title = "MainWindow" Height = "350" Width = "525">
    < Window.Resources >
        < SolidColorBrush x:Key = "bg" Color = "Yellow" />
    </Window.Resources >
    < WrapPanel >
        < Button Background = "{StaticResource bg}">A first button </Button >
        < Button Background = "{StaticResource bg}" > A second button </Button >
        < TextBox > I am a TextBox </TextBox >
        < TextBlock > I am a TextBlock </TextBlock >
    </WrapPanel >
</Window >
```

运行这段代码，两个按钮的背景都成为黄色，显示效果如图 1.10 所示。

资源绑定是数据绑定中比较简单的内容，下面将 TextBox 的文本属性绑定到 TextBlock 的文本属性上，实现代码如下。

```xml
< Window x:Class = "Button._0.MainWindow"
        xmlns = "http://schemas.microsoft.com/winfx/2006/xaml/presentation"
        xmlns:x = "http://schemas.microsoft.com/winfx/2006/xaml"
        Title = "MainWindow" Height = "350" Width = "525">
    < Window.Resources >
        < SolidColorBrush x:Key = "bg" Color = "Yellow" />
    </Window.Resources >
    < WrapPanel >
        < Button Background = "{StaticResource bg}">A first button </Button >
        < Button Background = "{StaticResource bg}" > A second button </Button >
        < TextBox x:Name = "txt">I am a TextBox </TextBox >
        < TextBlock Text = "{Binding ElementName = txt,Path = Text}" ></TextBlock >
    </WrapPanel >
</Window >
```

运行这段代码，在 TextBox 输入"This is a DataBinding example"，则 TextBlock 中的

内容也跟着改变,运行效果如图 1.11 所示。

图 1.10 Button 绑定资源

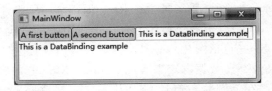

图 1.11 TextBox 与 TextBlock 的数据绑定

WPF 的可视化系统集成了 2D 矢量图、光栅图片、文本、音频、视频、动画和 3D 图形。所有这些特性都集中在一个引擎中,该引擎构建在 DirectX 之上,这样可以通过显卡硬件提升软件的性能。

2. 绘制 2D 图形

通过使用这种集成机制,绘制 2D 图形。设计方案是:在最外层使用< DockPanel >布局,在该布局中加入< Rectangle >,并使用< VisualBrush >对象让其重复显示的方式来填充,Viewport 和 TileMode 两个属性实现内容重复多次,代码如下。

```
< Window x:Class = "Button._0.MainWindow"
        xmlns = "http://schemas.microsoft.com/winfx/2006/xaml/presentation"
        xmlns:x = "http://schemas.microsoft.com/winfx/2006/xaml"
        Title = "MainWindow" Height = "350" Width = "525">
    < Window.Resources >
        < SolidColorBrush x:Key = "bg" Color = "Yellow" />
    </Window.Resources>
    < DockPanel >
        < WrapPanel x:Name = "wp" DockPanel.Dock = "Top">
            < Button Background = "{StaticResource bg}">A first button </Button>
            < Button Background = "{StaticResource bg}">A second button </Button>
            < TextBox x:Name = "txt">I am a text box </TextBox>
            < TextBlock Text = "{Binding ElementName = txt,Path = Text}"></TextBlock >
        </WrapPanel >
        < Rectangle Margin = "5" >
            < Rectangle.Fill >
                < VisualBrush Visual = "{Binding ElementName = wp}"
                    Viewport = "0,0,.5,.2"
                    TileMode = "Tile" >
                </VisualBrush >
            </Rectangle.Fill >
        </Rectangle >
    </DockPanel >
</Window >
```

运行这段代码,编辑文本框中的内容为"控件组建 2D 图形",矩形框中的内容也随之改变。运行效果如图 1.12 所示。

3. 绘制 3D 图形

下面绘制 3D 图形。首先创建 3D 场景,需从以下 5 个方面进行设计:模型(形状)、

图 1.12 控件组建 2D 图形

材质(贴图到模型)、相机(观看位置)、光源(照亮目标)、可视化区域(呈现场景的区域)。此处的材质用 VisualBrush,更为详细的内容将在后续章节中进行详述,实现 3D 图形的代码如下。

```
< Window x:Class = "Button._0.MainWindow"
        xmlns = "http://schemas.microsoft.com/winfx/2006/xaml/presentation"
        xmlns:x = "http://schemas.microsoft.com/winfx/2006/xaml"
        Title = "MainWindow" Height = "350" Width = "525">
    < Window.Resources >
        < SolidColorBrush x:Key = "bg" Color = "Yellow" />
    </Window.Resources >
    < DockPanel >
        < WrapPanel x:Name = "wp" DockPanel.Dock = "Top">
            < Button Background = "{StaticResource bg}">A first button </Button >
            < Button Background = "{StaticResource bg}" > A second button </Button >
            < TextBox x:Name = "txt">I am a text box </TextBox >
            < TextBlock Text = "{Binding ElementName = txt,Path = Text}"></TextBlock >
        </WrapPanel >
        < Viewport3D >
            < Viewport3D.Camera >
                < PerspectiveCamera
                    LookDirection = " - .7, - .8, - 1"
                    Position = "3.8,4,4"
                    FieldOfView = "17"
                    UpDirection = "0,1,0">
                </PerspectiveCamera >
            </Viewport3D.Camera >
            < ModelVisual3D >
                < ModelVisual3D.Content >
                    < Model3DGroup >
                        < PointLight
                            Position = "3.8,4,4"
                            Color = "White"
                            Range = "7"
                            ConstantAttenuation = "1.0 "/>
```

```xml
            < GeometryModel3D >
                < GeometryModel3D.Geometry >
                    < MeshGeometry3D
TextureCoordinates = "0,0 1,0 0,-1 1,-1 0,0 1,0 0,-1 0,0"
Positions = "0,0,0 1,0,0 0,1,0 1,1,0 0,1,-1 1,1,-1 1,1,-1 1,0,-1"
TriangleIndices = "0,1,2 3,2,1 4,2,3 5,4,3 6,3,1 7,6,1"/>
                </GeometryModel3D.Geometry >
                < GeometryModel3D.Material >
                    < DiffuseMaterial >
                        < DiffuseMaterial.Brush >
                            < VisualBrush
                                Viewport = "0,0,.5,.25"
                                TileMode = "Tile"
                                Visual = "{Binding ElementName = wp}"/>
                        </DiffuseMaterial.Brush >
                    </DiffuseMaterial >
                </GeometryModel3D.Material >
            </GeometryModel3D >
        </Model3DGroup >
      </ModelVisual3D.Content >
    </ModelVisual3D >
  </Viewport3D >
</DockPanel >
</Window >
```

运行这段代码,编辑文本框中的内容为"这是一个 3D 模型",显示效果如图 1.13 所示。

图 1.13 控件作材质组建 3D 模型

4. 动画

到目前为止，所有的内容都是静止的。WPF 支持动画，动画是根据时间来调整属性值的。再加入旋转变换可让 3D 模型动起来。在 3D 模型代码的基础上，加入旋转变换功能，代码如下，省略与绘制 3D 模型相同的代码。

```xml
<!-- 略 -->
    <GeometryModel3D>
        <GeometryModel3D.Transform>
            <RotateTransform3D
                CenterX=".5"
                CenterY=".5"
                CenterZ="-.5">
                <RotateTransform3D.Rotation>
                    <AxisAngleRotation3D
                        x:Name="rotation"
                        Axis="0,1,0"
                        Angle="0"/>
                </RotateTransform3D.Rotation>
            </RotateTransform3D>
        </GeometryModel3D.Transform>
<!-- 保留 Window 代码部分 -->
```

开始定义自己的动画，引入 <DoubleAnimation>，加入新的功能：在 3 秒内，完成一次从 −30° 角到 30° 角的旋转，重复该动作，新增功能代码如下，省略与绘制 3D 图形相同的代码。

```xml
<!-- 略 -->
    <Window.Triggers>
        <EventTrigger RoutedEvent="FrameworkElement.Loaded">
            <EventTrigger.Actions>
                <BeginStoryboard>
                    <Storyboard>
                        <DoubleAnimation
                            From="-30"
                            To="30"
                            Storyboard.TargetName="rotation"
                            Storyboard.TargetProperty="Angle"
                            AutoReverse="True"
                            Duration="0:0:3"
                            RepeatBehavior="Forever"/>
                    </Storyboard>
                </BeginStoryboard>
            </EventTrigger.Actions>
        </EventTrigger>
    </Window.Triggers>
<!-- 保留 Window 代码部分 -->
```

运行上述代码，动画效果如图 1.14 所示。

5. 设置样式

使用样式可以把一组属性应用到一个或多个控件上，实现一致的主题风格。

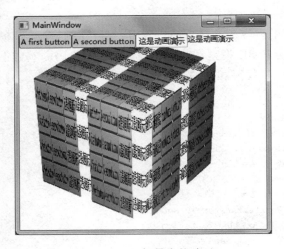

图 1.14　3D 场景中的动画

定义样式,设计方案:把背景设置移入样式中,代替上述代码中每个按钮指向的资源,将样式的键指向 Button 类型,则样式可用于窗口中所有的按钮。设置样式实现的功能代码如下。

```xml
<!-- 略 -->
<Window.Resources>
    <SolidColorBrush x:Key = "bg" Color = "Yellow" />
    <Style x:Key = "{x:Type Button}" TargetType = "{x:Type Button}">
        <Setter Property = "Background" Value = "{StaticResource bg }" />
        <Setter Property = "Template">
            <Setter.Value>
                <ControlTemplate TargetType = "{x:Type Button}">
                    <Grid>
                        <Ellipse StrokeThickness = "4">
                            <Ellipse.Stroke>
                                <LinearGradientBrush>
                                    <GradientStop Offset = "0" Color = "White" />
                                    <GradientStop Offset = "1" Color = "Green" />
                                </LinearGradientBrush>
                            </Ellipse.Stroke>
                            <Ellipse.Fill>
                                <LinearGradientBrush>
                                    <GradientStop Offset = "0" Color = "silver"/>
                                    <GradientStop Offset = "1" Color = "white"/>
                                </LinearGradientBrush>
                            </Ellipse.Fill>
                        </Ellipse>
                        <ContentPresenter
                            Margin = "10"
                            HorizontalAlignment = "Center"
                            VerticalAlignment = "Center" />
                    </Grid>
                </ControlTemplate>
            </Setter.Value>
        </Setter>
```

```
            </Style>
        </Window.Resources>
<!-- 保留 Window 代码部分 -->
```

运行设置样式功能代码,显示效果如图 1.15 所示。两个按钮的外观呈椭圆状,椭圆线框运用了从白到绿的渐变色。

图 1.15　样式实现椭圆按钮

1.4　WPF 体系结构

WPF 是多层的体系结构。在顶层,应用程序和一个完全由托管的 C#代码编写的一组高层服务进行交互。在底层执行将.NET 对象转换为 DirectX 任务。而这任务是在后台由一个名为 Milcore.dll 的低级的非托管组件完成的。Milcore.dll 是以非托管代码方式实现的。

1.4.1　WPF 运行机制

WPF 的核心是一个与分辨率无关的基于向量的呈现引擎,可以充分利用现代图形设备的优势,提高程序开发的效率。WPF 和公共语言运行环境(Common Language Runtime,CLR)的完全集成,确保 CLR 提供的类型安全、跨平台等特性。WPF 运行机制如图 1.16 所示,它包含 WPF 的主要组件。图中的黄色部分(PresentationFramework、PresentationCore 和 Milcore)是 WPF 的组件部分。在这些组件中,Milcore 是以非托管代码编写的,是 WPF 隐藏于 CLR 之下的核心驱动组件,目的是与 DirectX 的紧密集成,在实际应用程序开发中不太可能访问到，这里不再进一步介绍。

由 WPF 运行机制可知，WPF 中的所有显示都是通过 DirectX 引擎完成的，还要求对内存和执行进行精确控制，实现高效硬件管理及软件呈现。WPF 运行在 CLR 之上的，以托管代码的方式公开应用程序编程接口（Application Programming Interface，API），并提供自身的程序模型和类库（PresentationFramework 和 PresentationCore）。

WPF 体系结构中一个重要原理就是基于属性，WPF 提供的类库、操作方式等都尽可能地使用属性，而不是方法或事件。因为属性是声明性的，比方法和事件更加容易指定对象的意图（这在 XAML 中相当重要），所以属性系统是 WPF 体系结构中一个重要的组成部分。在 WPF 属性系统中，属性可以被继承和监视，当属性被更改时，属性联系的双方都被通知。由于被继承，因此子元素可以感受到父元素属性的变化，例如，父窗体

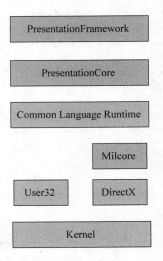

图 1.16　WPF 运行机制

的窗体大小属性被更改时，会自动通知到子窗体，并同步刷新界面。WPF 属性系统的根本是 System.Windows.DependencyObject 类型，它是 WPF 属性系统的基类。

在 WPF 中，System.Windows.Media.Visual 类型提供界面元素的显示支持，它用于生成一个可视化树，树中每个元素都包含特定的绘制和实现能力，从而将需要的数据显示到界面。另外，Visual 类还将托管的 WPF 组件和非托管的 Milcore 组件链接到一起，通过在屏幕上定义一个矩形显示区域来提供显示框架，从而将可视化树（可视化树在本节稍后部分介绍）中各元素呈现到屏幕上。

1.4.2　WPF 类层次结构

WPF 中大部分类都是从 UIElement、FrameworkElement、ContentElement、FrameworkContentElement 4 个类派生而来的，这 4 个类称为基元素类。其中，UIElement 是主要类，它是从 Visual 类派生而来的，适用于支持大型数据模型的元素，这些元素用于矩形屏幕区域的区域内呈现和布局，在该区域内，使用内容模型可以让不同的元素进行组合。

在 WPF 中，改变了传统 Windows 应用程序窗体中相对位置计算的布局模式，在 WPF 中的布局就像网页上元素的布局，显得更加灵活。图 1.17 为 WPF 类层次结构。

在此，对 WPF 类层次结构中的类做简要说明。

（1）System.Threading.DispatcherObject 类，继承此类，用户界面中的每个元素都可以检查代码是否在正确的线程上运行。

（2）System.Windows.DependencyObject 类，提供对依赖属性和附加属性的支持。

（3）System.Windows.Media.Visual 类，Visual 实际上是 WPF 组合系统的入口点。Visual 是托管 API 和非托管 Milcore 这两个子系统之间的连接点。每个元素本质都是一个 Visual 对象。

（4）System.Windows.UIElement 类，提供 WPF 本质特征的支持，如布局、输入、焦点及事件。

（5）System.Windows.FrameworkElement 类，提供数据绑定、动画及样式的支持。

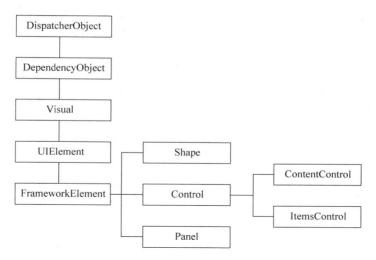

图 1.17　WPF 类层次结构

（6）System.Windows.Shapes.Shape 类，基本形状类继承于此类。
（7）System.Windows.Controls.Control 类，为字体、前背景色及模板提供支持。
（8）System.Windows.Controls.ContentControl 类，所有具有单一内容类控件的基类。
（9）System.Windows.Controls.ItemsControl 类，所有显示选项集合控件的基类。
（10）System.Windows.Controls.Panel 类，所有布局窗口的基类。

1.4.3　WPF 的可视化树与逻辑树

当人们看到一些设计风格新颖的网站时，可以借助浏览器自带的 Inspector 工具或插件方便地浏览网站布局结构及逻辑。如果是 WPF 开发的应用程序，不仅可以获得页面布局结构，还可以使用 WPF Inspector 工具来了解应用程序的架构方式。下面来介绍如何使用 WPF Inspector 查看 WPF 中的可视化树与逻辑树。

在 WPF 应用程序运行后，再启动 WPF Inspector，如图 1.18 所示。

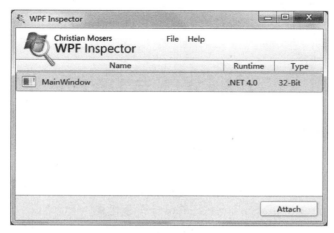

图 1.18　WPF Inspector 启动界面

WPF 编程基础

在此处运行 1.3 节设置样式中的"样式实现椭圆按钮"程序，再启动 WPF Inspector，单击 Attach 按钮，截取 Visual Tree，如图 1.19 所示。选择 Logic Tree 选项卡，得到的逻辑树如图 1.20 所示。

图 1.19 可视化树

图 1.20 逻辑树

在该例的 UI 中，窗口逻辑树的根结点 Window 下面有一个子结点 DockPanel。而 DockPanel 有两个子结点 WrapPanel 和 Viewport3D。WrapPanel 下有 4 个子结点：两个 Button、一个 TextBox 和一个 TextBlock。逻辑树始终存在于 WPF 的 UI 中，不管 UI 是用 XAML 编写还是用代码编写。WPF 的每个方面（属性、事件、资源等）都是依赖于逻辑树的。

可视化树是逻辑树的一种扩展。逻辑树的每个结点都被分解为它们的核心可视化组件。逻辑树的结点对用户而言基本是一个黑盒。而可视化树不同，它暴露了可视化的实现细节。可视化树比逻辑树全面，显示了每一层 UI 对象，可以对每一个可视 UI 对象进行控制，也可直接修改布局容器，需要显示特殊效果时，可以调用 System.Windows.LogicalTreeHelper 这个类来实现。

1.5 WPF 与 UWP

微软于 2006 年发布 WPF，截至 2016 年初，微软在我国已陆续推出了 UWP（Universal Windows Platform）应用，不过初始版本很简陋。那么微软的 UWP 与 WPF 两者之间又有什么关联呢？

UWP 即 Windows 10 中的 Universal Windows Platform 简称，即 Windows 通用应用平台，在 Windows 10 Mobile/Surface（Windows 平板电脑）/PC/Xbox/HoloLens 等平台上运行，UWP 不同于传统 PC 上的 EXE 应用，它与只适用于手机端的 APP 有着本质的区别。由它的名字就可以知道，它并不是为某一个终端而设计的，而是可以在所有 Windows 10 设备上运行的。下面将两者放在一起简要讨论。

（1）WPF 已经是比较成熟的技术，而 UWP 支持多种设备。

（2）WPF 所有的操作都不依赖于 GDI(Graphics Device Interface，图形设备界面)和 GDI+，而是间接依赖于强大的 DirectX，这就意味着通过 WPF 可以做出以前用 WinForm 无法想象的视觉效果，包括 3D 效果的应用程序。而 UWP 作为 Windows 通用应用平台，随着其应用产品的开发，也将慢慢凸显出其优势。

（3）WPF 彻底把程序架构、业务逻辑和用户界面(UI)分离开。WPF 引擎把 XAML 描述的 UI 元素解释为相应的.NET 对象，从而在应用程序创建相应的控件，UI 人员和程序人员均可对此控件进行编辑加载，实现用户界面和程序架构的彻底分离，而微软的 WinForm 做不到。WPF 是成熟的技术，是 UWP 开发的基础。

1.6 小　　结

本章从 Windows Form 托管对象模型出发，剖析了 WPF 的编程机制，了解了 XAML 编程模型与 XML 的关系。通过逐层深入的 Button 案例，认识了 WPF 平台特性，阐述了 WPF 的运行机制及类层次结构，从而厘清了 WPF 的体系结构。

习题与实验 1

1. 简述 WPF 的新特性及体系结构。

2. 分别用 HTML 和 XAML 编写代码，在 IE 中输出"This is my first WPF class!"，页面效果如图 1.21 与图 1.22 所示。

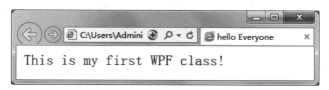

图 1.21　HTML 效果

图 1.22　XAML 效果

3. 设计 2D 和 3D 页面，效果如图 1.23 和图 1.24 所示，设计 3D 页面效果时，适当调整 3D 旋转角度。

4. 下载并安装 WPF Inspector 工具，观察第 3 题的可视化树与逻辑树。

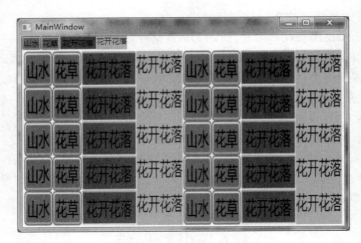

图1.23 2D效果图

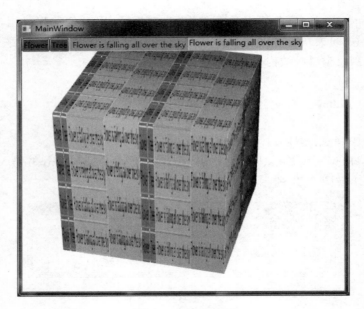

图1.24 3D效果图

第 2 章 XAML

XAML(eXtensible Application Markup Language,可扩展应用程序标记语言)是微软公司创建的一种新的描述性语言,用于搭建应用程序用户界面。XAML 实现了用户界面与业务逻辑完全分离。

在第 1 章中,对照了 HTML 与 XAML 两者的异同点。发现尽管 XAML 在元素的声明、程序样式的设置等语法结构上和 HTML 上有相似之处,但 XAML 有别于 HTML 最显著的特征是 xmlns 指令和标记扩展。其中,xmlns 指令需要在标记中关联名称空间。

XAML 并不是 HTML,而是基于 XML 的,是 WPF 的外在表现形式。而 HTML 只是一种标记语言,仅仅是用来为浏览器呈现页面内容。XAML 可以用来呈现信息和请求用户输入等基本的功能,它还支持动画和 3D 等高级特性。

XAML 是可扩展的,开发人员可以创建自定义的控件、元素和函数来扩展 XAML。由于 XAML 各元素在本质上就是 WPF 类的映射,因此开发人员可以很轻松地使用面向对象的技术对 XAML 元素进行扩展。也就是说,可以开发一些自定义控件和组合元素,用户界面设计人员和开发人员直接使用即可,软件真正做到了可复用性。

2.1 XAML 文档框架

一个简单的 WPF 应用程序 UI 页面,都能由 XAML 编写出来。下面创建 WPF 应用程序,命名为"Button._1",进入应用程序启动页面后,在解决方案管理器中双击文件名,打开 MainWindow.xaml,系统自动生成的代码如下。

```
<Window x:Class = "Button._1.MainWindow"
        xmlns = "http://schemas.microsoft.com/winfx/2006/xaml/presentation"
        xmlns:x = "http://schemas.microsoft.com/winfx/2006/xaml"
        Title = "MainWindow" Height = "350" Width = "525">
    <Grid>
    </Grid>
</Window>
```

在此首先了解 XAML 文件基本框架,这个文档包含一个顶级元素 Window 和一个 Grid 元素。Window 代表整个窗口,Grid 布局上可以放置所需控件。一个 XAML 文件只能有一个顶级元素。

WPF 中可以当顶级元素的还有 Page 元素和 Application 元素。Page 用于可导航的应用程序;Application 用于定义应用程序的资源和启动设置。

分析 Window 开始标签,依照自上而下的顺序,它包含一个类名、两个 xmlns 指令关联的名称空间和 3 个属性(Title、Height 和 Width)。

2.1.1　XAML 文档结构

在 XAML 文件基本框架上,为应用程序添加设计要求：在 Button 上放置一个 Image 和一个 TextBox。在< Grid >与</Grid >之间,添加 XAML 代码如下。

```
<!-- 保留 Window 代码部分 -->
<Grid>
<Button>
    <Button.Content>
        <StackPanel Orientation = "Vertical">
            <Image Source = "happyface.jpg" Width = "150" Height = "150"></Image>
            <TextBox Text = " Smile" VerticalAlignment = " Center" HorizontalAlignment =
            "Center">
            </TextBox>
        </StackPanel>
    </Button.Content>
</Button>
</Grid>
</Window>
```

运行上述代码,显示效果如图 2.1 所示。分析代码的逻辑结构可知,Window 为顶级元素,在 Grid 布局下有一个 Button；在 Button 下使用 StackPanel 布局后,又加入了 Image 和 TextBox。将本案例的 XAML 文件结构用图 2.2 表示。

图 2.1　XAML 笑脸界面

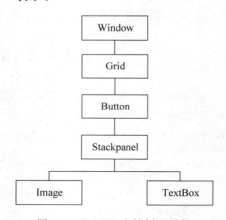

图 2.2　XAML 文档树形结构

应用程序的 UI 在用户看来是一个平面结构,与传统设计理念不同的是,XAML 是树形逻辑结构。当然实现 Button 上放置 Image 和 TextBox 的 UI 有很多方式,但是不管哪种方式,文档框架都是树形结构。

2.1.2　基础语法

XAML 中最基本的语法元素与 XML 类似,下面回顾一下 XML 中的标签、属性和内容的语法。

1. 标签

标签通常是以<>开始,以</>结束的,一个标签的声明通常表示一个对象。如< Window ></Window >、< Grid ></Grid >分别定义了一个窗体对象及一个 Grid 对象,标签定义有以下两种常用写法。

(1) 非自闭合标签:如< Window ></Window >、< Grid ></Grid >,标签要成对出现。

(2) 自闭合标签:如< Window />、< Grid/>、< Button/>。

这种自闭合标签用于无内容情况下,可以让代码看上去更简洁,当然,正常情况下 Window 及 Grid 都是有内容的。

2. 属性

属性通常以键值对形式出现,形式如下。

Attribute = Value

例如,< Window ></Window >标签中的"Title = "MainWindow" Height = "350" Width = "525"",等号左边表示 Window 标签的属性,等号右边表示该属性的值。

3. 内容

一组标签对之间夹杂的文本或其他标签都称为这个标签之间的内容。此处 Window 标签的内容就是一对< Grid ></Grid >标签。

2.2 XAML 中的属性

本节继续使用 Button._1 代码,根据需要还会不断丰富其功能,以便配合讲解 XAML 中的各属性。

2.2.1 简单属性

XAML 是声明性语言,XAML 编译器给每个标签创建一个与标签对应的对象,再对其属性初始化。所以,每个标签是要先声明对象,再为对象赋初值。赋值方法有两种:一种是 XAML 中的字符串简单赋值,另外一种是在后台 CS 代码中,使用属性元素进行复杂赋值。

1. 字符串简单赋值

XAML 使用标签定义 UI 元素,每一个标签对应.NET Framework 类库的一个控件类。通过设置标签的 Attribute(属性),实现标签对应的控件对象 Property(属性)赋值。

```
< Grid >
    < Rectangle x:Name = "rectangle" Fill = "Blue" Margin = "344,112,56,125"></Rectangle >
</Grid >
```

2. 属性元素复杂赋值

在 CS 代码中,创建对象,并设置它们的属性。代码如下。

```
public partial class MainWindow : Window
{
    public MainWindow()
    {
        InitializeComponent();
```

```
        SolidColorBrush scb = new SolidColorBrush();
        scb.Color = Colors.Yellow;
        this.rectangle.Fill = scb;
    }
}
```

两种赋值方法相对照，这里需解释 Fill 属性的用法。因为 Fill 类型是 Brush，但是 Brush 是一个抽象类，由于抽象类不能实例化，所以只能用抽象类的子类实例赋值。Brush 的派生类简要说明如下。

- SolidColorBrush：使用纯 Color 绘制区域；
- LinearGradientBrush：使用线性渐变绘制区域；
- RadialGradientBrush：使用径向渐变绘制区域；
- ImageBrush：使用图像（由 ImageSource 对象表示）绘制区域；
- DrawingBrush：使用 Drawing 绘制区域。绘图可能包含向量和位图对象；
- VisualBrush：使用 Visual 对象绘制区域，使用 VisualBrush 可以将内容从应用程序的一个部分复制到另一个区域，这在创建反射效果和放大局部屏幕时会非常有用。

2.2.2 复杂属性

使用 Button._1，加入新的设计需求：Grid 的背景使用渐变色，渐变方案是：由红色到黄色，再由黄色到绿色。

要实现该功能就需要对 Grid.Background 中的 LinearGradientBrush 进行设置，代码如下。

```
<!-- 保留 Window 代码部分 -->
<Grid>
    <Grid.Background>
        <LinearGradientBrush EndPoint = "0.5,1" StartPoint = "0.5,0">
            <GradientStop Color = "Red" Offset = "0" />
            <GradientStop Color = "Yellow" Offset = "0.6" />
            <GradientStop Color = "Green" Offset = "1" />
        </LinearGradientBrush>
    </Grid.Background>
<!-- 保留 Button.1 后续代码部分 -->
```

运行上述代码，显示效果如图 2.3 所示。代码中的 Grid.Background 便是复杂属性。下面介绍程序中用到的属性。

在上节的学习中，已知 LinearGradientBrush 的功能是线性渐变笔刷填充线性渐变颜色到当前区域，故可搭配两种或两种以上的颜色。它的重要属性有 GradientStop（倾斜点）、Color（渐变颜色）、Offset（偏移量）、StartPoint（起点坐标）、EndPoint（终点坐标）。

LinearGradientBrush 的渐变色是由多个

图 2.3 Grid 渐变效果

GradientStop 组成的,其中 Color 和 Offset 分别表示渐变色值和颜色起始位置。Offset 宽度是整个绘制区域,取值范围为 0~1.0。

2.2.3 附加属性

在上述代码的基础上,加入新的设计需求:Grid 布局中,加入一个紫色矩形。显示效果如图 2.4 所示。

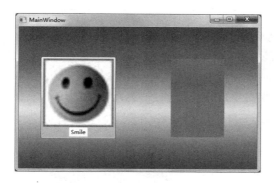

图 2.4 加入紫色矩形后呈现效果

由图 2.4 可知,需要为对 Grid 新增加 1 列,对< Grid.ColumnDefinitions >属性设置代码如下。

```
<!-- 保留上述代码部分 -->
</Grid.Background>
        <Grid.ColumnDefinitions>
            <ColumnDefinition></ColumnDefinition>
            <ColumnDefinition></ColumnDefinition>
        </Grid.ColumnDefinitions>
```

要将< Rectangle >置于 Grid 的第 1 列(在 Grid 中的行与列的起始值都是从 0 开始),紫色(Fill="Violet"),代码如下。

```
<!-- 保留上述代码部分 -->
    </Button>
        <Rectangle x:Name="Rec" Grid.Column="1" Fill="Violet" Margin="70" /></Grid>
</Window>
```

查阅上述两段代码中用到的属性,< Grid.ColumnDefinitions >是复杂属性。"Grid.Column="1""位于< Rectangle >内部,这就是附加属性。

这时读者可能有疑问,两个属性都是"Grid."这种形式,为什么称谓不同,因为 ColumnDefinitions 这一属性隶属于 Grid,它完全可以视为 Grid 的子标签,而"Grid.Column="1""是在< Rectangle >内部,用来设置 Rectangle 标签的。它并非真正的属性,被转为调用 Get.SetColumn()方法来实现。再进一步观察,会发现附加属性用在控件布局里面。

2.2.4 处理特殊字符与空白

在 XAML 中,可以让一些特殊字符作为元素的内容出现,例如,上述示例中的 Smile 要变成<Smile>显示,需要修改 TextBox 标签中的 Text="<smile>"。

表 2.1 列出了常用的特殊字符,并给出它在 XAML 标签中的处理方式。

表 2.1 XAML 中特殊字符处理方式

特殊字符	处理方式	特殊字符	处理方式
<(小于号)	&lt;	&(与字符)	&amp;
>(大于号)	&gt;	"(英文引号)	&auot;

在 XAML 中处理特殊字符并不难,但还存在一个关键问题——空白的处理。在默认情况下,XAML 忽略所有的空白,这就意味着空格、Tab 键、硬回车、带有多个空格的长字符串被转换为单个空格。有时放在文本中的空格,希望输出,例如,文本框中输出带尖括号及空格的字符,如<Smile>,则需要修改代码如下,显示效果如图 2.5 所示。

<TextBox Name = "Txt" Text = "< smile >" xml:space = "preserve" />

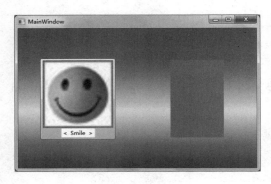

图 2.5 特殊字符与留白效果

在保留空格的元素(TextBox)中使用"xml:space="preserve"",Text 属性中的空格被保留输出。事实上"xml:space="preserve""特性是 XML 标准的一部分。

2.3 XAML 名称空间

本节将进一步介绍 XAML 文件根标记中的两个默认 XAML 名称空间映射及其用途,同时还介绍如何生成类似的映射,便于用在代码中或单独的程序集中定义的元素。

2.3.1 名称空间的作用

开发语言会将常用功能以类的形式封装,开发人员根据业务需求,也会封装符合自身业务需求的类,有序组织这些类。这样,一方面,便于开发人员准确调用;另一方面,编译器可以有效地识别具有相同命名的类,就引入了名称空间。名称空间是通过类似树形结构来组织各种类,是一种较为有效的类名排列方式。

XAML 和.NET 其他语言一样,也是通过名称空间有效地组织 XAML 内部的相关元素类,但是 XAML 的名称空间与.NET 中的名称空间不是一对一的映射关系,而是一对多映射。

2.3.2 默认名称空间

在上一节 XAML 文档结构中,了解到系统默认的两个名称空间的代码如下。

```
xmlns = "http://schemas.microsoft.com/winfx/2006/xaml/presentation"
xmlns:x = "http://schemas.microsoft.com/winfx/2006/xaml"
```

XAML 名称空间是 XML 名称空间概念的扩展。xmlns 是 XML-Namespace 的缩写,看起来类似"网址",但在 IE 中无法打开,所以它并不是网址。它是遵循 XAML 解析器标准的命名规则,是声明程序集和.NET 名称空间的引用。

其中,"xmlns = "http://schemas.microsoft.com/winfx/2006/xaml/presentation""与.NET 的名称空间中的 System.Windows 下的类相对应,包含 XAML 基本的布局和控件。

带 x 的"xmlns:x = "http://schemas.microsoft.com/winfx/2006/xaml""对应一些 XAML 语法和编译相关的 CLR 名称空间。

为了避免概念上的混淆,在此,对 C♯、CLR 与.NET 做简要说明。在人们做开发时,常会使用 C♯ 语言,它是.NET 的核心开发语言;.NET(.NET Framework)是一个运行时平台;CLR(Common Language Runtime)是公共语言运行库,与 Java 虚拟机一样是一个运行时环境,它负责资源管理(内存分配和垃圾收集等),确保应用和底层操作系统之间必要的分离。正是因为有了 CLR,所以.NET 成为一个跨语言的集成开发平台。

2.3.3 名称空间中的标记扩展

标记扩展是 XAML 的重要特性,它放在一对花括号{}中,使用类似对象标签的方式来精确解析。

现在增加新的设计需求:矩形框的颜色为黄色。简单的方法是 Fill = "Yellow",但是在此,用标记扩展来实现。先设置窗体资源,代码如下。

```
< Window.Resources >
    < SolidColorBrush x:Key = "bg" Color = "Yellow" />
</Window.Resources >
```

再修改 Rectangle 的 Fill,代码如下。

```
< Rectangle x:Name = "Rec" Grid.Column = "1" Fill = "{StaticResource bg}" Margin = "70" />
```

运行代码后,得到如图 2.6 所示的效果。Fill 使用标记扩展来赋值。

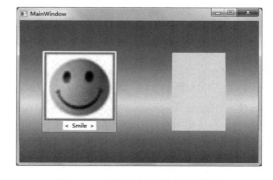

图 2.6 名称空间中的标记扩展

表2.2是 XAML 内置 XAML 特性名称空间指令及含义,并使用标记扩展语法示例说明。

表2.2 XAML 内置 XAML 特性

XAML 名称空间指令	含 义	示 例
x:Array	创建 CLR 数组	< x:ArrayType="{x:Type Button}"> < Button/> < Button/> </x:Array >
x:Class	定义的类名	< Window 　x:Class="MainWindow">… </Window >
x:ClassModifier	定义类型的模式 Public、Internal	< Window… 　x:Class="MainWindow">… 　x:ClassModifier="Public"… </Window >
x:Code	XAML 中内嵌代码	< x:Code > 　Public void SomeMethod(){} </x:Code >
x:Key	包含在字典中的元素键	< Window.Resources > 　< SolidColorBrush x:Key="bg"…/> </Window.Resources >
x:Name	指定元素的编程名	< Rectangle x:Name="Rec" …/>
x:Null	创建一个空值	< TextBox Text="{x:Null}">
x:Static	静态字段和属性值	< Button background="{Static…}"/>
x:Type	提供 CLR 类型	< ControlTemplate > 　TargetType="{x:Type Button}" </ControlTemplate >
x:TypeArguments	为实例化泛型类型指定泛型类型的参数	< object x:Class=" namespace.classname " x:TypeArguments="{x:Type type1}[,{x:Type type2},{x:Type type3},…]"></object >
x:XData	XAML 指令元素	主要用作 XmlDataProvider 的子对象或 XmlDataProvider.XmlSerializer 属性的子对象

2.4 类型转换器

XAML 的属性元素赋值形式为 Attribute=Value。此处的 Value 是一个 String 类型,但在后台 CS 代码中,对象的属性(Property)类型不一定是 String 类型。为了解决 XAML 与 CS 代码中的数据类型不统一,使用 TypeConverter 类。这个类的功能是:将值类型转换为其他类型,如可以把字符串转换成对象。

现在又在设计中新增需求:在 CS 中创建 Teacher 类,Teacher 中有 Name 和 Student 两个属性,其中 Name 类型是 String,Student 类型是 Teacher;XAML 中的 Button 按下后,

弹出一个消息框,消息框中的显示内容为 I am your sun。其中,消息框调用方式为 Teacher.Student.Name。

根据设计要求,首先在后台 CS 中创建 Teacher 类,代码如下。

```
public class Teacher
    {
        public string Name { get;set; }
        public Teacher Student { get; set; }
    }
```

在 XAML 中变更的代码如下。

```
<Window x:Class = "Button._1.MainWindow"
    xmlns = "http://schemas.microsoft.com/winfx/2006/xaml/presentation"
    xmlns:x = "http://schemas.microsoft.com/winfx/2006/xaml"
    xmlns:local = "clr-namespace:Button._1"
    Title = "MainWindow" Height = "350" Width = "525">
<Window.Resources>
    <local:Teacher x:Key = "teacher" Student = "I am your sun"/>
</Window.Resources>
<!-- 保留 Window 代码部分 -->
    <Button Name = "BtnMes" Height = "180" Width = "160" Click = "BtnMes_Clicked">
<!-- 保留 Window 代码部分 -->
</Window>
```

因为 Button 上放着图片,双击 Button,系统无法跳转到 Click 事件。在此,建议操作方式:在事件"Click＝"BtnMes_Clicked""上右击,弹出如图 2.7 所示的快捷菜单,选择"导航到事件处理程序"选项,编辑 Click 事件代码如下。

```
private void BtnMes_Clicked(object sender, RoutedEventArgs e)
{
    Teacher t = (Teacher)this.FindResource("teacher");
    MessageBox.Show(t.Student.Name);
}
```

运行代码,单击,抛出异常,报错"Resource Reference Key Not Found Exception"。找不到调用资源,出错位置为"MessageBox.Show(t.Student.Name)"。分析出错原因是:XAML 中的属性值类型无法转换成 CS 中的对象类型。解决问题的方案是通过 TypeConverter 派生用户自定义类,重载该类的 ConvertFrom 方法。该方法中有一个 Value,它就是在 XAML 中进行设置的,再把这个 Value 转换成用户期望的数据类型。下面创建一个将 String 转换成 Teacher 的用户自定义类,将这个类命名为 StringToTeacherTypeConverter,CS 代码如下。

```
using System.ComponentModel;          //Add new Namespace
namespace Button._1
{
    public class StringToTeacherTypeConverter:TypeConverter
    {
        public override object ConvertFrom(ITypeDescriptorContext context, System.Globalization.CultureInfo culture, object value)
```

```
        {
            if (value is string)
            {
                Teacher t = new Teacher();
                t.Name = value as string;
                return t;
            }
            return base.ConvertFrom(context, culture, value);
        }
    }
}
```

如果现在就运行代码，还是会报错，出错类型和上面相同。在 Teacher 类前，编写代码如下。

```
using System.ComponentModel;                //Add new Namespace
namespace Button._1
{
    [TypeConverterAttribute(typeof(StringToTeacherTypeConverter))]
    public class Teacher
    {
        public string Name { get;set; }
        public Teacher Student { get; set; }
    }
}
```

现在运行代码，单击 Button，弹出消息框，显示页面效果如图 2.8 所示。

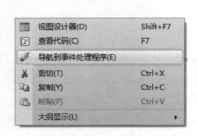

图 2.7 导航到事件处理程序

图 2.8 TypeConverter 实现

此处可以把 TypeConverterAttribute 中的 Attribute 省去，因为特征类在使用时可省略 Attribute。

2.5　导入程序集

在项目开发时，一个项目由多个模块构成，模块之间可以相互调用。在用 Visual Studio 2010 创建的解决方案下，每个项目都可以独立编译，独立编译的结果就得到一个程序集。

常见的程序集包括 exe(可执行文件)和 dll(动态链接库)。这里说的"导入程序集"是指 dll 类型。程序集要放在合适的名称空间下,名称空间解决不同模块类名相同问题。

在 XAML 中导入程序集。设用户自定义的程序集名为 UserLibrary.dll,它有 Comm 和 Cont 两个名称空间,在 XAML 中引用这两个名称空间的语法格式如下。

```
xmlns:映射名 = "clr-namespace:名称空间;assembly = 程序集名"
```

接下来根据语法格式,在程序集 UserLibrary.dll 中的两个名称空间对应的 XAML 引用为:

```
xmlns:comm = "clr-namespace:Comm;assembly = UserLibrary"
xmlns:cont = "clr-namespace:Cont;assembly = UserLibrary"
```

在 XAML 文档添加引用后,就可以使用名称空间中的类,语法格式如下。

```
<映射名:类名>…</映射名:类名>
<comm:类名></comm:类名>
<cont:类名></cont:类名>
```

本节的例子只是简单引入,后面会详细说明。

2.6 小 结

本章从 XAML 文档框架开始,让读者认识到 UI 的平面结构对应着 XAML 文档的树形结构。XAML 是可扩展的应用程序声明式语言,它是基于 XML 的,所以 XAML 中的标签、属性、内容语法结构与 XML 有相似之处。本章案例从 Button 上放置 Image、TextBox 的例子开始,通过分层叠加式案例,逐步地讲解 XAML 的复杂属性、附加属性、xmlns 指令和名称空间中的标记扩展,并使用 TypeConverter 类把字符串转换成对象。还讲解了项目开发时,在 XAML 导入程序集的语法。倘若在学习中,有些内容不懂,可以跳过,当遇到具体应用后,回过头来阅读,会恍然大悟。对 XAML 语法知识有所了解以后,便可以用它来创建用户界面了。

习题与实验 2

1. 简述 XAML 中的简单属性、复杂属性和附加属性。
2. 在 XAML 页面定义两个矩形,左边的矩形填充成紫色,右边的矩形填充为渐变色,渐变从蓝色到紫色再到红色。外部的 Grid 布局背景色用米黄色(beige),页面显示效果如图 2.9 所示。
3. 设计软件登录界面,界面效果如图 2.10 所示。该登录界面设计要求:界面中按钮背景色(参照标记扩展部分)绑定同种颜色,当单击"登录"按钮时,可以弹出"欢迎登录"对话框。

图 2.9　紫色渐变矩形

图 2.10　登录界面

第 3 章 布局

WPF 布局是通过面板（Panel）对页面元素进行全面规划和安排。简单地说，就是把一些控件有条理地摆放在界面上合适的位置。在应用程序界面设计中，合理的元素布局至关重要，它可以方便用户操作，并用清晰的页面逻辑呈现用户信息。如果内置布局控件不能满足需要，用户还可以创建自定义的布局元素。

3.1 布局原则

WinForm 的布局是采用基于坐标的方式，当窗口内容发生变化时，里面的控件不会随之动态调整，用户体验不够好。而 WPF 采用了基于流的布局方式，像 Web 开发模式。流式布局特点是：所有的元素总是默认地自动向左上角靠近，在设计时，通过控制元素相对位置的方式使其达到预计的效果，即元素的位置依赖于相邻元素的位置和尺寸。

3.1.1 合成布局模型

WPF 的合成布局模型是用来满足广泛的应用场景布局，允许某种布局控件被嵌套在其他布局控件中。合成布局模型通过布局契约来实现子控件和父布局控件间的通信问题。布局契约包括两种设计思想，即根据内容调整尺寸和两段布局。

1. 根据内容调整尺寸

根据内容调整尺寸，即每个控件都根据内容来确定控件大小，这个设计思想应用于 UI 中的所有控件。例如，窗口能够调整大小来适应它们内部的控件，文本框控件能调整尺寸来适应它内部的文本。当然每个元素会被询问其期望的尺寸大小，以确保根据内容调整尺寸的设计思想能够实施。

2. 两段布局

两段布局是指在两个完全不同的阶段来确定控件的最佳尺寸。在这两个阶段布局模型让父布局控件和子控件达成元素最后尺寸的约定。两个阶段分别是测量（Measure）和排列（Arrange）。测量阶段需要做的主要工作是：对整个 UI 页面的检测，并询问每个元素的期望尺寸（Desired Size），元素返回一个可用的尺寸（Available Size），当所有的元素都被询问并测量好以后，就进入到排列阶段。在排列阶段，父元素通知每个子元素的实际尺寸（Actual Size）和位置。

在两段布局中，父元素和子元素需要协商出需要的尺寸大小，涉及可用尺寸、期望尺寸、实际尺寸，在此辨析 3 个尺寸。其中，可用尺寸是测量阶段的初始约束值，即父元素愿意给子元素的最大空间值；期望尺寸是子元素想要的尺寸；实际尺寸是父元素分配给子元素的最终尺寸。这 3 个尺寸要符合下面的不等式条件。

Desired Size≤Actual Size≤Available Size

了解 WPF 合成布局模型，学习 WPF 布局机制，才能理解合成布局模型的来龙去脉，在页面布局时做到得心应手。

3.1.2 布局机制

WPF 界面上的每个元素的边界框尺寸和排列是 WPF 自动计算出来的。通过 WPF 合成布局模型的学习，了解 WPF 渲染布局的过程中，执行测量（Measure）和排列（Arrange）两个步骤。在布局机制中，详细分解在 WPF 布局的不同阶段，后台类的调用过程。在测量阶段，布局容器遍历所有子元素，并询问子元素所期望的尺寸；在排列阶段，布局控件在合适的位置放置子元素，并设置元素的最终尺寸；这是一个递归的过程，界面中任何一个容器元素都会被遍历到。

因为面板可以嵌套，所以处理过程是递归的，布局（Layout）处理的过程如图 3.1 所示。

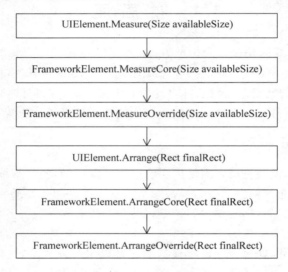

图 3.1 WPF 布局原理

在此，简要说明 WPF 布局处理过程。由 WPF 框架可知，所有 UI 元素的根元素是 UIElement 类型，在 UIElement 中定义了一些基本的关于 UI 显示的属性（如 Clip 和 Visibility）。在 UIElement.Measure(Size availableSize)方法执行阶段，就是对这些基本属性做评估，获得适合的 Size。同样，FrameworkElement.MeasureCore(Size availableSize)方法评估时，在 FrameworkElement 中定义且有可能影响 UI 布局的属性，得出更适合的 Size。这个 Size 将被传递给 FrameworkElement.MeasureOverride(Size availableSize)方法。WPF 提供的 Panel 类型（如 Grid）中就会重写该方法来处理，处理完后将得到一个系统期望的 Size(称为 DesiredSize)。布局系统将按照这个 Size 来显示该 Element，测量（Measure）阶段结束。Size 确定后，把 Size 包装为 Rect 实例，传递给 UIElement.Arrange(Rect finalRect)，进行排列（Arrange）处理。根据 Size 值，Arrange 方法为元素创建边界框，边框打包到 Rect 实例，传给 FrameworkElement.ArrangeCore(Rect finalRect)方法。ArrangeCore

将继续评估 DesiredSize,计算边界留白(Margin,Padding)等信息,获得 ArrangeSize,并传给 FrameworkElement.ArrangeOverride(Size finalSize)。这个方法也是可重写的,WPF 提供的 Panel 类型会重写该方法来处理,最终获得 finalSize。当 finalSize 确定后,ArrangeOverride 执行完毕,控制权回到 ArrangeCore 方法,ArrangeCore 把该 Element 放到它的边界框中。到此,该 Element 的 Layout 处理完成。

3.1.3 布局通用属性

所有的 WPF 布局面板都由 System.Windows.Controls.Panel 抽象类派生。Panel 就是所有布局元素的基类,用于放置和排列 WPF 元素,这个抽象类包含 3 个公共属性:Background、Children 和 IsItemHost(IsItemHost 标志着控件是不是类似 TreeView 和 ListView 这样的控件)。布局容器内的子元素对自身的对齐方式(HorizontalAlignment、VerticalAlignment)、宽度(Width)、高度(Height)、四周间隙(Margin)等有一定的决定权。子元素可以设置自身的布局属性来调整自己的位置和大小。表 3.1 给出了 WPF 布局面板中的通用属性。

表 3.1 WPF 布局面板中的通用属性

属 性 名 称	表 征 意 义
Background	布局面板背景着色,在响应鼠标事件时,该值非空(含透明)
Children	布局面板中存储的条目集合,条目还可以含更多的条目
IsItemHost	布尔值,是否为由 ItemsControl 生成的 UI 项的容器
HorizontalAlignment	水平对齐方式,有 Center、Left、Right、Stretch 属性值可选
VerticalAlignment	垂直对齐方式,有 Center、Left、Right、Stretch 属性值可选
MinWidth/MinHeight	最小宽度尺寸/最小高度尺寸,默认单位是像素
MaxWidth/MaxHeight	最大宽度尺寸/最大高度尺寸,默认单位是像素
Width/Height	宽度尺寸/高度尺寸,默认单位是像素
Margin	在元素周围空白尺寸,默认单位是像素

3.2 布 局 面 板

WPF 的布局面板实现基本的布局,布局面板类型有 Canvas、DockPanel、StackPanel、WrapPanel、Grid 和 UniformGrid。在这一节中,不涉及 Grid,因为 Grid 灵活,适用场合颇多,需独立讲解。表 3.2 列出了布局面板的类型及其使用规则。

表 3.2 布局面板的类型及其使用规则

布局面板类型	使 用 规 则
Canvas	画布。通过 Top、Left、Right 和 Bottom 4 个属性将子元素定位
DockPanel	停靠面板。让子元素停靠在整个面板的某一条边上,然后拉伸元素以填满全部宽度或高度。类似于 Windows Form 编程中控件的 Dock 属性
StackPanel	堆式面板。子元素按照声明的先后顺序,自上往下或从左往右摆放

续表

布局面板类型	使用规则
WrapPanel	自动折行面板。子元素按照声明的先后顺序,从左往右摆放,摆满一行后,自动折行。与 HTML 中的流式布局相似
Grid	网格。通过定义行高和列宽来调整子元素。类似于 HTML 中的 Table
UniformGrid	简化网格布局。每个单元格具有相同的大小,自动创建相同的行列数

3.2.1 Canvas

Canvas(画布)是 WPF 中最简单的布局控件,是用于存储控件的容器,不会自动调整内部元素的排列及大小,它仅支持用显式坐标定位控件。可以使用 Left、Top、Right 和 Bottom 附加属性在 Canvas 中定位控件。实质上,在选择每个控件停靠角时,附加属性的值是作为外边距使用的。如果一个控件没有使用任何附加属性,它会被放在 Canvas 的左上方(等同于设置 Left 和 Top 为 0)。

Canvas 实例 1,设计要求:将 5 个 Button 按钮分别放在画布的左上角、右上角、左下角、右下角和中心位置。使用 XAML 代码如下。

```
< Canvas Width = "200" Height = "100" Background = "Beige">
    < Button Content = "Left, Top" Canvas.Left = "4" Canvas.Top = "4"/>
    < Button Content = "Right, Top" Canvas.Right = "4" Canvas.Top = "4"/>
    < Button Content = "Left, Bottom" Canvas.Left = "4" Canvas.Bottom = "4"/>
    < Button Content = "Right, Bottom" Canvas.Right = "4" Canvas.Bottom = "4"/>
    < Button Content = "Center" Canvas.Left = "75" Canvas.Top = "38"/>
</Canvas >
```

运行上述代码后,显示页面效果如图 3.2 所示。"画布"上的元素附加属性 Canvas.Left、Canvas.Top、Canvas.Right 和 Canvas.Bottom 来完成内部子元素(Button)的定位。这里的 Canvas.Left 和 Canvas.Top 属性与 Windows Form 窗体控件的 Left 和 Top 属性类似,Canvas.Left 属性表示距离"画布"左边的距离,而 Canvas.Top 属性表示距离"画布"顶部的距离。

Canvas 实例 2,设计要求:分别将黄色、粉红色和蓝紫色 3 个 Rectangle 放到画布上,使用 XAML 代码如下。

```
< Canvas >
    < Rectangle Canvas.ZIndex = "3" Width = "60" Height = "60" Canvas.Top = "30"
        Canvas.Left = "30" Fill = "BlueViolet"/>
    < Rectangle Canvas.ZIndex = "1" Width = "60" Height = "60" Canvas.Top = "50"
        Canvas.Left = "50" Fill = "Yellow"/>
    < Rectangle Canvas.ZIndex = "2" Width = "60" Height = "60" Canvas.Top = "70"
        Canvas.Left = "70" Fill = "Pink"/>
</Canvas >
```

运行上述代码后,显示页面效果如图 3.3 所示。"画布"上的元素附加属性 Canvas.ZIndex 设置不同颜色的 Rectangle 声明的先后顺序,从视觉效果上看,后声明的蓝紫色矩形将先声明的矩形框覆盖,用专业术语则是 Canvas.ZIndex 设置重叠深度。

图 3.2　Canvas 附加属性定位

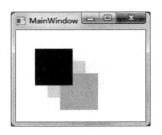

图 3.3　Canvas.ZIndex 设置重叠深度

3.2.2　DockPanel

　　DockPanel(停靠面板)让子元素停靠在整个面板的某一条边上。当多个子元素停靠在相同的边时,根据声明的先后顺序依次停靠,系统默认最后一个声明子元素来添满剩余空间。打开 Windows 计算机窗口,如图 3.4 所示,该窗口是利用 DockPanel 的常见布局方式。

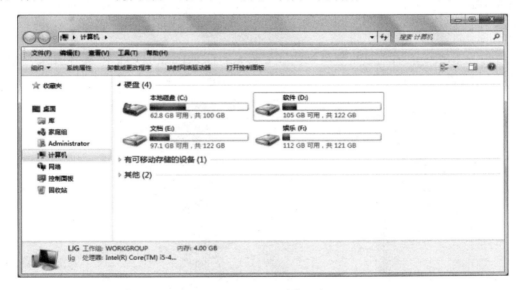

图 3.4　Windows 计算机窗口

　　现将 Windows 计算机窗口分解后,它是由菜单栏、工具条、文件夹列表框、详细信息列表框和状态栏组成的,如图 3.5 所示。

　　现在使用停靠面板来创建 Windows 计算机窗口的基本结构,在此用 Button 来表示所有的子元素。XAML 代码如下。

```
<DockPanel>
    <Button DockPanel.Dock = "Top">菜单栏区域</Button>
    <Button DockPanel.Dock = "Top">工具条区域</Button>
    <Button DockPanel.Dock = "Bottom">状态栏区域</Button>
    <Button DockPanel.Dock = "Left">文件夹区域</Button>
    <Button DockPanel.Dock = "Right">详细信息列表区域</Button>
</DockPanel>
```

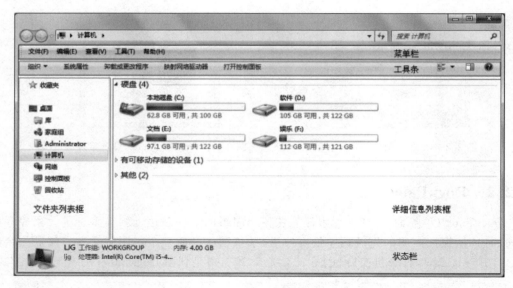

图 3.5　Windows 计算机窗口分解

运行上述代码后,页面显示效果如图 3.6 所示。DockPanel.Dock 作为 Button 的附加属性,用来设置上(Top)、下(Bottom)、左(Left)、右(Right)的停靠位置。

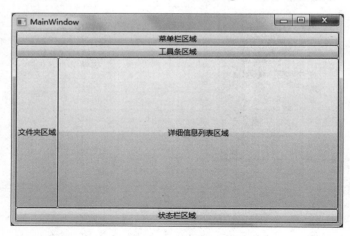

图 3.6　DockPanel 构建窗口常用布局

下面通过调整 Button 声明的顺序来改变页面布局结构。XAML 代码如下。

```
<DockPanel>
    <Button DockPanel.Dock = "Top">菜单栏区域</Button>
    <Button DockPanel.Dock = "Left">文件夹区域</Button>
    <Button DockPanel.Dock = "Top">工具条区域</Button>
    <Button DockPanel.Dock = "Bottom">状态栏区域</Button>
    <Button DockPanel.Dock = "Right">详细信息列表区域</Button>
</DockPanel>
```

运行上述代码后,页面显示效果如图 3.7 所示。读者留意会发现,在文件夹区域与详细信息列表区域之间没有分隔条。具有分隔条的 WPF 控件是 GridSplitter,它依赖于 Grid 控件,后续章节将对其进行讨论。

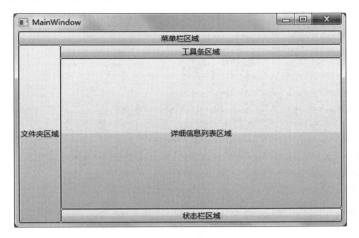

图 3.7　调整 Button 声明顺序改变布局结构

3.2.3　StackPanel

StackPanel(堆式面板)将子元素按照声明的先后顺序堆在一起。设置 Orientation 属性值确定以水平或垂直方向堆放。

在 StackPanel 中,根据子元素的最大尺寸来确定最佳尺寸。设计 StackPanel 实例,页面显示效果如图 3.8 所示。

设计要求:在最外层是边框环绕,内部一个 StackPanel 中再嵌套两个 StackPanel,子元素用 Button,XAML 代码如下。

```
< Border BorderBrush = "Black" BorderThickness = "1"
        HorizontalAlignment = "Center" VerticalAlignment = "Center">
    < StackPanel Height = "154" Width = "221">
        < StackPanel Margin = "2" Background = "Yellow" Orientation = "Vertical">
            < Button Content = "One" Margin = "1"/>
            < Button Content = "Two" Margin = "1"/>
            < Button Content = "Three" Margin = "1"/>
            < Button Content = "Four" Margin = "1"/>
            < Button Content = "Five" Margin = "1"/>
        </StackPanel>
        < StackPanel Margin = "2" Background = "pink" Orientation = "Horizontal">
            < Button Content = "One" Margin = "3"/>
            < Button Content = "Two" Margin = "3"/>
            < Button Content = "Three" Margin = "3"/>
            < Button Content = "Four" Margin = "3"/>
            < Button Content = "Five" Margin = "3"/>
        </StackPanel>
    </StackPanel>
</Border>
```

StackPanel 会根据方向使用无限宽度和高度来测量子控件,当有其他布局同时出现时,会影响布局的整体风格。本例中,通过最外层的 Border 对 StackPanel 进行约束。

StackPanel 适合局部页面布局,接下来用它做一个简单的用户搜索页面。页面效果如图 3.9 所示。XAML 代码如下。

```
<StackPanel Background = "AliceBlue">
    <TextBlock Margin = "4">Look for:</TextBlock>
    <ComboBox Margin = "4"></ComboBox>
    <TextBlock Margin = "4">Filtered by:</TextBlock>
    <ComboBox Margin = "4"></ComboBox>
    <Button Margin = "4,5">Search</Button>
    <CheckBox Margin = "4">Search in titles only</CheckBox>
    <CheckBox Margin = "4">Search in Keyword</CheckBox>
</StackPanel>
```

图 3.9 是由两个 TextBlock、两个 ComboBox、两个 CheckBox 和一个 Button 构成的页面,设置 Margin 属性让页面保持清爽。

图 3.8 3 个 StackPanel 嵌套

图 3.9 用户搜索页面

3.2.4 WrapPanel

WrapPanel(自动折行面板)允许任意多的子元素按照声明的先后顺序,从左往右摆放,摆满一行后,自动折行。折行面板的经典的例子就是工具条布局。

默认情况下,WrapPanel 根据子元素的内容,自动调整控件的大小。也可以设置 ItemWidth 和 ItemHeight 属性来约束子元素的宽度和高度。WrapPanel 布局 10 个 Button,如图 3.10 与图 3.11 所示。

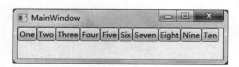

图 3.10 WrapPanel 布局 10 个 Button

图 3.11 折行的 Button

由图 3.10 可知,XAML 代码如下。

```
<WrapPanel Background = "AliceBlue">
    <Button>One</Button>
```

```
    <Button>Two</Button>
    <Button>Three</Button>
    <Button>Four</Button>
    <Button>Five</Button>
    <Button>Six</Button>
    <Button>Seven</Button>
    <Button>Eight</Button>
    <Button>Nine</Button>
    <Button>Ten</Button>
</WrapPanel>
```

把图 3.10 中的窗口宽度变窄,调整到一行放 5 个 Button,图 3.10 会变成图 3.11 的效果。

3.2.5 UniformGrid

UniformGrid(简化网格布局)隐含在 System.Windows.Controls.Primitives 名称空间中。每个单元格具有相同的大小。在使用 UniformGrid 布局时,需要设定行值与列值。下面使用 UniformGrid 对 6 个 Button 布局,XAML 代码如下。

```
<Border BorderBrush="Black" BorderThickness="1"
        HorizontalAlignment="Center" VerticalAlignment="Center">
    <UniformGrid Rows="3" Columns="2" Background="Green">
        <Button Content="One" Margin="1"/>
        <Button Content="Two" Margin="1"/>
        <Button Content="Three" Margin="1"/>
        <Button Content="Four" Margin="1"/>
        <Button Content="Five" Margin="1"/>
        <Button Content="Six" Margin="1"/>
    </UniformGrid>
</Border>
```

运行上述代码,页面显示效果如图 3.12 所示。UniformGrid 布局是绿色背景,6 个 Button,按照 3 行 2 列生成大小相同的单元格。

在使用 UniformGrid 布局时,如果只设定列值,那么行值等于子元素的数目除以列值;如果只设定行值,那么列值等于子元素的数目除以行值。接下来,使用 UniformGrid 布局 6 个 Button 为 2 行 3 列,XAML 代码如下。

```
<Border BorderBrush="Black" BorderThickness="1"
        HorizontalAlignment="Center" VerticalAlignment="Center">
    <UniformGrid Columns="3" Background="Green">
        <Button Content="One" Margin="1"/>
        <Button Content="Two" Margin="1"/>
        <Button Content="Three" Margin="1"/>
        <Button Content="Four" Margin="1"/>
        <Button Content="Five" Margin="1"/>
        <Button Content="Six" Margin="1"/>
    </UniformGrid>
</Border>
```

运行上述代码,页面显示效果如图 3.13 所示。代码中只设置了"Columns="3"",

UniformGrid 布局 6 个 Button，系统行值等于 6 除于 2，自动生成 2 行 3 列的 UniformGrid 网格布局。

图 3.12　网格 3×2 的 UniformGrid

图 3.13　网格 2×3 的 UniformGrid

使用 UniformGrid 布局的网格宽度是所有子元素中的最大宽度值；高度是所有子元素中的最大高度值。用 UniformGrid 布局 3×3 的网格，XAML 代码如下。

```
<UniformGrid Columns = "3" Rows = "3" Background = "AliceBlue" >
    < Button Content = "One" Margin = "1"/>
    < Button Content = "Two" Margin = "1"/>
    < Button Content = "Three" Margin = "1"/>
    < Button Content = "Four" Margin = "1"/>
    < Button Content = "Five" Margin = "1"/>
    < Button Content = "Six" Margin = "1"/>
    < Button Content = "Seven" Margin = "1"/>
</UniformGrid >
```

运行上述代码，页面显示效果如图 3.14 所示。由图 3.14 可知，当代码中给出的单元格数大于子元素的数目，依次摆放后，余下的单元格为空。

修改上述 XAML 代码，将"Rows＝"3""改成"Rows＝"2""，运行代码，页面显示效果如图 3.15 所示。可知，当系统提供的单元格数小于子元素的数目，UniformGrid 把最后一个子元素放到了边界外。

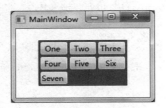

图 3.14　网格 3×3 的 UniformGrid

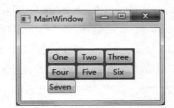

图 3.15　子元素大于网格单元数的 UniformGrid

3.3　Grid

尽管 UniformGrid 能够布局统一单元格，但是很多布局中需要构建单元格大小不等，具有跨越式单元格（Span Cells）和空白列等功能。而 Grid 能实现上述这些功能，它是一个使用灵活、能构建复杂 UI 的布局控件。

Grid 最简单的用法是通过设置 RowDefinitions 和 ColumnDefinitions 属性定义单元格的总数。其中，RowDefinitions 指行数，ColumnDefinitions 指列数。添加子元素时，使用

Grid.Row 和 Grid.Column 附加属性设定子元素在单元格中的位置。接下来,用 Grid 布局 6 个 Button,单元格排列成 3 行 2 列。XAML 代码如下。

```xml
<Grid Background="green">
    <Grid.RowDefinitions>
        <RowDefinition></RowDefinition>
        <RowDefinition></RowDefinition>
        <RowDefinition></RowDefinition>
    </Grid.RowDefinitions>
    <Grid.ColumnDefinitions>
        <ColumnDefinition></ColumnDefinition>
        <ColumnDefinition></ColumnDefinition>
    </Grid.ColumnDefinitions>
    <Button Grid.Row="0" Grid.Column="0" Margin="2">One</Button>
    <Button Grid.Row="0" Grid.Column="1" Margin="2">Two</Button>
    <Button Grid.Row="1" Grid.Column="0" Margin="2">Three</Button>
    <Button Grid.Row="1" Grid.Column="1" Margin="2">Four</Button>
    <Button Grid.Row="2" Grid.Column="0" Margin="2">Five</Button>
    <Button Grid.Row="2" Grid.Column="1" Margin="2">Six</Button>
</Grid>
```

运行上述代码,页面显示效果如图 3.16 所示。当调整窗口大小时,Button 大小随着窗口的变化而变化。

在很多应用程序中都会有登录页面的设计,在此,使用 Grid 布局用户登录页面,XAML 代码如下。

```xml
<Grid Background="AliceBlue">
    <Grid.ColumnDefinitions>
        <ColumnDefinition></ColumnDefinition>
        <ColumnDefinition></ColumnDefinition>
    </Grid.ColumnDefinitions>
    <Grid.RowDefinitions>
        <RowDefinition></RowDefinition>
        <RowDefinition></RowDefinition>
        <RowDefinition></RowDefinition>
    </Grid.RowDefinitions>
    <TextBlock Text="Username: " VerticalAlignment="Center" FontSize="20" HorizontalAlignment="Center" FontFamily="Times New Roman">
    </TextBlock>
    <TextBlock Text="Password: " Grid.Row="1" VerticalAlignment="Center" HorizontalAlignment="Center" FontSize="20" FontFamily="Times New Roman">
    </TextBlock>
    <TextBox Grid.Column="1" Margin="25" FontSize="20"></TextBox>
    <TextBox Grid.Column="1" Grid.Row="1" Margin="25" FontSize="20"></TextBox>
    <Button Grid.Row="2" Margin="25" Content="Login" FontSize="20" FontFamily="Times New Roman" Background="LightBlue">
    </Button>
    <Button Grid.Row="2" Margin="25" Content="Cancel" Grid.Column="1" FontSize="20" FontFamily="Times New Roman" />
</Grid>
```

运行上述代码,页面显示效果如图 3.17 所示。该登录页面含有两个 TextBlock、两个 TextBox 和两个 Button,共 6 个子元素。当窗口大小变化时,页面子元素会随着窗口的大小动态调整。

图 3.16 3×2 的 Grid

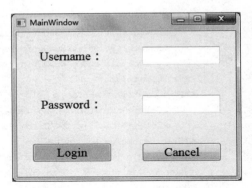

图 3.17 Grid 布局登录界面

上面的两个例子,显示了基本的 Grid 布局网格的用法,但并没有充分实现 Grid 的重要特征。下面从结构中分离布局、尺寸模型、共享尺寸组、跨越行和列与 GridSplitter 等方面来演示 Grid 的多种用法。

3.3.1 从结构中分离布局

前面讲到的 WPF 布局面板,都是通过改变元素的声明顺序来改变布局结构,因为声明顺序不同,所以布局结构也会不同。WPF 除 Grid 以外的布局面板,首先,将布局面板融入可视化树中,便于调用布局算法。然后,根据声明子控件声明的顺序,设定布局结构。

Grid 能从结构中分离布局,指的是不依赖子元素的声明顺序来改变布局结构。因为子元素可以通过两个附加属性进行定位。下面对 6 个 Button 通过设置 Grid.Row 和 Grid.Column 两个附加属性值,让页面显示与 Button 的定义顺序相反。XAML 代码如下。

```xml
<!-- 略与上例代码相同部分 -->
    < Button Grid.Row = "2" Grid.Column = "1" Margin = "2"> One </Button >
    < Button Grid.Row = "2" Grid.Column = "0" Margin = "2" > Two </Button >
    < Button Grid.Row = "1" Grid.Column = "1" Margin = "2" > Three </Button >
    < Button Grid.Row = "1" Grid.Column = "0" Margin = "2"> Four </Button >
    < Button Grid.Row = "0" Grid.Column = "1" Margin = "2" > Five </Button >
    < Button Grid.Row = "0" Grid.Column = "0" Margin = "2"> Six </Button >
</Grid>
```

运行上述代码后,页面显示效果如图 3.18 所示。6 个 Button 布局效果与 Button 定义顺序相反,是通过设置附加属性 Grid.Row 和 Grid.Column 的值来改变 Button 的输出顺序。

接下来选用 Uniform Grid 布局 6 个 Button,XAML 代码如下。

```xml
< Border BorderBrush = "Black" BorderThickness = "1"
        HorizontalAlignment = "Center" VerticalAlignment = "Center">
    < UniformGrid Columns = "2" Background = "Green" >
```

```
            < Button Content = "Six" Margin = "1"/>
            < Button Content = "Five" Margin = "1"/>
            < Button Content = "Four" Margin = "1"/>
            < Button Content = "Three" Margin = "1"/>
            < Button Content = "Two" Margin = "1"/>
            < Button Content = "One" Margin = "1"/>
        </UniformGrid>
</Border>
```

运行上述代码,页面显示效果如图 3.19 所示,在使用 UniformGrid 布局时,只能通过改变元素定义顺序改变布局结构。

图 3.18 设置 Grid 附加属性调整布局结构　　图 3.19 Uniform Grid 改变元素顺序改变布局结构

Grid 布局页面时,在不影响代码的情况下,通过设置 Grid.Row 和 Grid.Column 两个附加属性值,调整布局结构,这种从结构中分离布局使 Grid 能够创建控件结构。

3.3.2 尺寸模型

在 Canvas 布局中,要使用元素绝对值来分割空间;而 Grid 引入了百分比尺寸,也就是列或行的高度能设置为星花(*)单位。星花允许行和列在按照内容尺寸或绝对尺寸分配空间后,占用网格空间的一个百分比值。

接下来理解 * 的用法,XAML 代码如下。

```
< Grid Background = "green">
    < Grid.RowDefinitions >
        < RowDefinition Height = "50"></RowDefinition >
        < RowDefinition Height = "1 * "></RowDefinition >
        < RowDefinition Height = "2 * "></RowDefinition >
    </Grid.RowDefinitions >
    < Grid.ColumnDefinitions >
        < ColumnDefinition Width = "80"></ColumnDefinition >
        < ColumnDefinition ></ColumnDefinition >
    </Grid.ColumnDefinitions >
    < Button Grid.Row = "0" Grid.Column = "0" Margin = "2"> One </Button >
    < Button Grid.Row = "0" Grid.Column = "1" Margin = "2"> Two </Button >
    < Button Grid.Row = "1" Grid.Column = "0" Margin = "2"> Three </Button >
    < Button Grid.Row = "1" Grid.Column = "1" Margin = "2"> Four </Button >
    < Button Grid.Row = "2" Grid.Column = "0" Margin = "2"> Five </Button >
    < Button Grid.Row = "2" Grid.Column = "1" Margin = "2"> Six </Button >
</Grid>
```

运行上述代码后,页面显示效果如图 3.20 所示。第 0 行第 0 列的 Button One 的高度(Height="50")是 50 像素,宽度(Width="80")是 80 像素。Button Three 所在行占余下空间的 1/3,Button Five 所在行占余下空间的 2/3。还使用了 Grid.Row、Grid.Column 附加属性把子元素定位到单元格。在此,需要声明的是,Grid 的行和列都是从 0 开始计数的,如果没有指定子元素的行列值,则子元素默认位于第 0 行第 0 列。

将上述代码中的 Height 与 Width 全改成星花(*)单位的 XAML 代码如下。

```
<Grid Background = "green">
    <Grid.RowDefinitions>
        <RowDefinition Height = "1*"></RowDefinition>
        <RowDefinition Height = "2*"></RowDefinition>
        <RowDefinition Height = "3*"></RowDefinition>
    </Grid.RowDefinitions>
    <Grid.ColumnDefinitions>
        <ColumnDefinition Width = "1*"></ColumnDefinition>
        <ColumnDefinition Width = "1*"></ColumnDefinition>
    </Grid.ColumnDefinitions>
    <Button  Grid.Row = "0" Grid.Column = "0"   Margin = "2">One</Button>
    <Button  Grid.Row = "0" Grid.Column = "1"   Margin = "2">Two</Button>
    <Button  Grid.Row = "1" Grid.Column = "0"   Margin = "2">Three</Button>
    <Button  Grid.Row = "1" Grid.Column = "1"   Margin = "2">Four</Button>
    <Button  Grid.Row = "2" Grid.Column = "0"   Margin = "2">Five</Button>
    <Button  Grid.Row = "2" Grid.Column = "1"   Margin = "2">Six</Button>
</Grid>
```

运行上述代码后,页面显示效果如图 3.21 所示。第 0 行第 0 列的高度(Height="1*")是加权百分比,表示占总数(1+2+3=6)中的 1 份,即六分之一;第 1 行第 1 列的高度占六分之二,第 2 行第 2 列的高度占六分之三。同理,可求 Width 宽度所占的百分比,各为二分之一。

图 3.20 设置 Grid 附加属性调整布局结构

图 3.21 Uniform Grid 改变元素顺序改变布局结构

Height 属性和 Width 属性可以被设置成绝对值、百分比和 Auto。当采用绝对值时,像素是默认单位,可省略。

设置行高或者列宽时,除了可以使用像素作为单位外,还能使用厘米(Centimeter)、英寸(Inch)和点(Point)。它们与像素之间的换算关系如下。

- 1cm=(96/2.54)pixel;
- 1in=96pixel;

- 1pt＝(96/72)pixel。

3.3.3 共享尺寸组

共享尺寸组是使用 Grid 布局时,位于同一列的子元素具有相同的尺寸大小。若在一个窗体中,允许一组控件具有相同的尺寸,尺寸大小由最宽的那个控件来决定。

设置共享尺寸,需要两个步骤。首先,在某一列或行上设置 SharedSizeGroup 属性;在控件上设置"IsSharedSizeScope="True""。

为了演示这个效果,对位于同一窗口的 4 个 Button 实现共享尺寸。XAML 代码如下。

```
<StackPanel>
    <Grid IsSharedSizeScope = "True">
        <Grid.RowDefinitions>
            <RowDefinition></RowDefinition>
            <RowDefinition></RowDefinition>
        </Grid.RowDefinitions>
        <Grid.ColumnDefinitions>
            <ColumnDefinition Width = "Auto" SharedSizeGroup = "a"/>
            <ColumnDefinition Width = "Auto" SharedSizeGroup = "a"/>
        </Grid.ColumnDefinitions>
        <Button Grid.Row = "0" Grid.Column = "0">One</Button>
        <Button Grid.Row = "0" Grid.Column = "1">Two</Button>
        <Button Grid.Row = "1" Grid.Column = "0">Three(which is longer)</Button>
        <Button Grid.Row = "1" Grid.Column = "1">Four</Button>
    </Grid>
</StackPanel>
```

运行上述代码后,页面显示效果如图 3.22 所示。尽管列中设置了 Width＝"Auto",但是 4 个 Button 全与最宽的那个 Button 保持一致,因为列上设置了共享尺寸分组(SharedSizeGroup＝"a"),并且在 Grid 中设置"IsSharedSizeScope＝"True""。

将代码中的"IsSharedSizeScope＝"True""和"SharedSizeGroup＝"a""去掉,运行后,页面显示效果如图 3.23 所示。Button(Two)与 Button(Four)不再与最宽的 Button 保持一致。

图 3.22 设置共享尺寸的 Grid

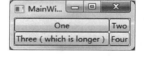

图 3.23 未设置共享尺寸的 Grid

3.3.4 跨越行和列

前面的例子,用 Grid.Row 和 Grid.Column 附加属性把子元素定位到单元格。还可以使用另外两个附加属性 Grid.RowSpan 和 Grid.ColumnSpan,让子元素跨越多个单元格。下面使用 Grid 设计计算器的页面,XAML 代码如下。

```
<Grid Background = "Green">
```

```xml
<Grid.RowDefinitions>
    <RowDefinition></RowDefinition>
    <RowDefinition></RowDefinition>
    <RowDefinition></RowDefinition>
    <RowDefinition></RowDefinition>
    <RowDefinition></RowDefinition>
    <RowDefinition></RowDefinition>
</Grid.RowDefinitions>
<Grid.ColumnDefinitions>
    <ColumnDefinition></ColumnDefinition>
    <ColumnDefinition></ColumnDefinition>
    <ColumnDefinition></ColumnDefinition>
    <ColumnDefinition></ColumnDefinition>
</Grid.ColumnDefinitions>
<TextBox Grid.Row="0" Grid.ColumnSpan="4" Margin="2,4,2,2"/>
<Button Grid.Row="1" Grid.Column="0" Margin="2">1</Button>
<Button Grid.Row="1" Grid.Column="1" Margin="2">2</Button>
<Button Grid.Row="1" Grid.Column="2" Margin="2">3</Button>
<Button Grid.Row="1" Grid.Column="3" Margin="2">+</Button>
<Button Grid.Row="2" Grid.Column="0" Margin="2">4</Button>
<Button Grid.Row="2" Grid.Column="1" Margin="2">5</Button>
<Button Grid.Row="2" Grid.Column="2" Margin="2">6</Button>
<Button Grid.Row="2" Grid.Column="3" Margin="2">-</Button>
<Button Grid.Row="3" Grid.Column="0" Margin="2">7</Button>
<Button Grid.Row="3" Grid.Column="1" Margin="2">8</Button>
<Button Grid.Row="3" Grid.Column="2" Margin="2">9</Button>
<Button Grid.Row="3" Grid.Column="3" Margin="2">*</Button>
<Button Grid.Row="4" Grid.ColumnSpan="2" Margin="2">0</Button>
<Button Grid.Row="4" Grid.Column="2" Margin="2">.</Button>
<Button Grid.Row="4" Grid.Column="3" Margin="2">/</Button>
<Button Grid.Row="5" Grid.ColumnSpan="2" Margin="2">Del</Button>
<Button Grid.Row="5" Grid.Column="2" Grid.ColumnSpan="2" Margin="2">=</Button>
</Grid>
```

运行上述代码,页面显示效果如图3.24所示。Grid布局6行4列网格。页面是由一个TextBox、17个Button构成。

其中,<TextBox Grid.Row="0" Grid.ColumnSpan="4" Margin="2,4,2,2"/>这条代码中的"Grid.Row="0""用来设置TextBox位于第0行;"Grid.ColumnSpan="4""设置TextBox占据4列,即从第0列~第3列;"Margin="2,4,2,2""设置TextBox与四周边界的左、上、右、下相距四周的间隙分别是2像素、4像素、2像素、2像素。

修改上述代码中的两条语句:

```xml
<Button Grid.Row="4" Grid.ColumnSpan="2" Margin="2">0</Button>
<Button Grid.Row="5" Grid.ColumnSpan="2" Margin="2">Del</Button>
```

修改后的语句:

```xml
<Button Grid.Row="4" Grid.RowSpan="2" Margin="2">0</Button>
<Button Grid.Row="4" Grid.RowSpan="2" Grid.Column="1" Margin="2">Del</Button>
```

运行修改后的代码,页面显示效果如图 3.25 所示。其中"Grid.RowSpan="2""实现跨越 2 行。

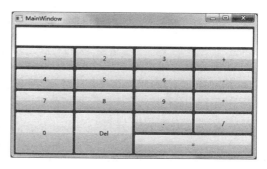

图 3.24 Grid 布局计算器页面　　　　　　图 3.25 设置附加属性的计算器页面

3.3.5　GridSplitter

GridSplitter 是网格分割线。它支持用户在运行时编辑行或列,当用户移动分割线时,还可以改变行高列宽。在很多聊天软件的页面布局都含有分割线。接下来,设计常用的"一上二下式"布局。该布局的上面是一个菜单栏,下面是两列网格,两列网格间是一条分割线。在布局中的子元素全部用 Button,XAML 代码如下。

```
<DockPanel>
    <Button Height = "20" DockPanel.Dock = "Top">Menu</Button>
    <Grid>
        <Grid.ColumnDefinitions>
            <ColumnDefinition Width = "3 * "/>
            <ColumnDefinition Width = "7 * "/>
        </Grid.ColumnDefinitions>
        <Button VerticalContentAlignment = "Center"
            HorizontalContentAlignment = "Center">Column1</Button>
        <GridSplitter Width = "3"></GridSplitter>
        <Button VerticalContentAlignment = "Center" Grid.Column = "1"
            HorizontalContentAlignment = "Center">Column2</Button>
    </Grid>
</DockPanel>
```

运行上述代码,页面显示效果如图 3.26 所示,当程序处在运行状态下,用户可以根据需要,移动 Column1 与 Column2 之间的 GridSplitter。

将图 3.26 页面中的 Column2 中加上一条垂直分割线,XAML 代码如下。

```
<DockPanel>
    <Button Height = "20" DockPanel.Dock = "Top">Menu</Button>
    <Grid>
        <Grid.ColumnDefinitions>
            <ColumnDefinition Width = "3 * "/>
            <ColumnDefinition Width = "7 * "/>
        </Grid.ColumnDefinitions>
```

```
                < Button VerticalContentAlignment = "Center"
                        HorizontalContentAlignment = "Center">Column1 </Button>
                < GridSplitter Width = "3"></GridSplitter>
                < Button VerticalContentAlignment = "Center" Grid.Column = "1"
                        HorizontalContentAlignment = "Center" > Column2 </Button>
                < Grid Name = "GridRightColumn" Grid.Row = "0" Grid.Column = "1">
                    < Grid.RowDefinitions >
                        < RowDefinition ></RowDefinition >
                        < RowDefinition ></RowDefinition >
                    </Grid.RowDefinitions >
                    < Button > Row1 </Button >
                    < GridSplitter Height = "3"   HorizontalAlignment = "Stretch"
                        VerticalAlignment = "Bottom"></GridSplitter >
                    < Button Grid.Row = "1"> Row2 </Button >
                </Grid >
            </Grid >
        </DockPanel >
</Window >
```

运行上述代码,页面显示效果如图 3.27 所示。分析布局结构发现,在 DockPanel 下是 Grid,Grid 下还有一个 Grid。其中,语句"< Grid Name = "GridRightColumn" Grid.Row = "0" Grid.Column="1">"指的是为第 2 个 Grid 命名,并使用了第一个 Grid 的附加属性定位;"< GridSplitter Height = "3" HorizontalAlignment = "Stretch" VerticalAlignment = "Bottom">"语句定义垂直分割线水平拉伸。

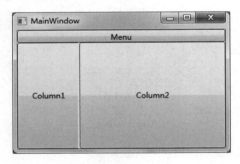

图 3.26 GridSplitter 实现水平分割

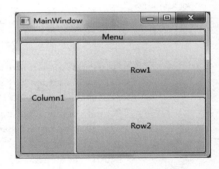

图 3.27 GridSplitter 实现水平＋垂直分割

3.4 小　　结

本章重点介绍了 WPF 布局原则以及布局面板中的 Canvas、DockPanel、StackPanel、WrapPanel 和 UniformGrid 的适用场合;详细介绍了 Grid 从结构中分离布局、尺寸模型、共享尺寸组和跨越行列等特征,并演示了 Grid 的多种用法。本章中的案例涉及 Windows 窗口页面、用户搜索页面、用户登录页面、计算器页面及常用布局。但是布局内容远不止这些,当了解更多的控件以后,可以做出个性化的布局。一个好的布局,能让 UI 根据用户的需求调整屏幕大小,窗口的大小,并根据内容来改变尺寸。在以后的章节中还会陆续补充布局的相关知识。

习题与实验 3

1. 简述流式布局的特点。

2. 分析如图 3.28 所示的 QQ 聊天页面，用 Grid 布局页面，页面中出现的元素用 Button 表示，其效果如图 3.29 所示。

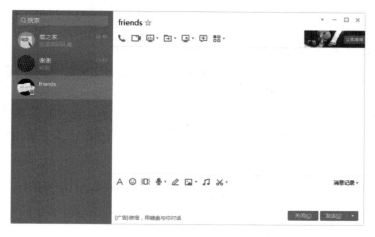

图 3.28　QQ 聊天页面

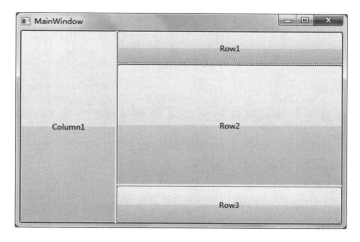

图 3.29　Grid 布局 QQ 聊天页面

操作提示：QQ 聊天页面的结构：先分为左右结构，右半部分又分成上、中、下结构，用 Grid 嵌套实现。在内部页面共用 3 根 GridSplitter 分割线。最下面的分割线，需要指定这个属性"Grid.Row＝"1""。

第 4 章

控件

第 3 章讲述了 WPF 布局原则及布局库中的各布局面板的用法。每个应用程序的 UI 都是由窗口及用户控件构成的。在 UI 工具包出现之前，就开始有控件。用户的界面也是由控件组成的。而本章中所介绍的控件都继承自 System.Windows.Control 类的元素。

尽管控件有自己默认的外观，但是完全可以通过属性设置控件的前景色或背景色来改变控件的外观。在深入学习 WPF 控件之前，先来了解 WPF 控件与众不同的新理念：内容模型和模板。

4.1 WPF 控件新理念

由第 1 章内容可知，WPF 提供了统一的编程模型、语言和框架，可以把 WPF 视为 UI 文档、媒体于一体的集成开发平台。在 WPF 之前的框架中，控件不灵活。以 Button 为例，用在不同场合的 Button，需要遵循元素合成（Element Composition）的原则。在以往的通过属性添加内容到按钮中，内容只能是字符串类型，而 WPF 中，Button 上可以放置一张图片，这就用到了"内容模型"这个新理念。

4.1.1 内容模型

WPF 使用大多数开发人员所熟悉的 CLR 类型系统。设置 Button 的内容，CS 代码如下。

```
Button b = new Button();
b.Content = "Hello WPF";
```

上述代码中 Button 的 Content 属性类型是 System.Object，而不是字符串。在 XAML 页面上，放置一个 Button 按钮，运行程序后，启动 WPF Inspector，生成 Button 的可视化树如图 4.1 所示。分析可视化树中的 Button 按钮可知，它是由 3 个元素合成的，如图 4.2 所示。

由图 4.2 可知，一个 Button 是由 ButtonChrome、ContentPresenter 和 TextBlock 3 个元素合成的。其中，ButtonChrome 用于显示按钮背景，TextBlock 用于显示基本文本的类型。下面重点介绍 ContentPresenter。

1. ContentPresenter

ContentPresenter 是内容模型的呈现器，用于显示 Content 属性中的数据，并生成一个可视化树，下面的 ContentPresenter 示例用于显示字符串和日期时间，后台 CS 代码如下。

第 4 章 控件

图 4.1 只有一个 Button 的可视化树

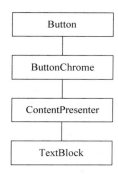

图 4.2 Button 控件的元素合成机制

```
public MainWindow()
{
    InitializeComponent();
    StackPanel panel = new StackPanel();
    ContentPresenter strPresenter = new ContentPresenter();
    strPresenter.Content = "Hello WPF";
    panel.Children.Add(strPresenter);
    ContentPresenter datePresenter = new ContentPresenter();
    datePresenter.Content = DateTime.Now;
    panel.Children.Add(datePresenter);
    ContentPresenter elementPresenter = new ContentPresenter();
    elementPresenter.Content = new Button();
    panel.Children.Add(elementPresenter);
    Content = panel;
}
```

运行上述代码,它的可视化树如图 4.3 所示,页面效果如图 4.4 所示。

图 4.3 ContentPresenter 对象显示不同类型数据的可视化树

图 4.4 ContentPresenter 对象显示数据的页面

转到 ContentPresenter 对象模型，查看定义，代码如下。

```
public class ContentPresenter : FrameworkElement
{       ...
    public object Content { get; set; }
    public string ContentSource { get; set; }
    public string ContentStringFormat { get; set; }
    public DataTemplate ContentTemplate { get; set; }
    public DataTemplateSelector ContentTemplateSelector { get; set; }
    public bool RecognizesAccessKey { get; set; }
    protected override Size ArrangeOverride(Size arrangeSize);
    protected virtual DataTemplate ChooseTemplate();
    protected override Size MeasureOverride(Size constraint);
    protected virtual void OnContentStringFormatChanged(string
            oldContentStringFormat, string newContentStringFormat);
    protected virtual void OnContentTemplateChanged(DataTemplate
            oldContentTemplate, DataTemplate newContentTemplate);
    protected virtual void OnContentTemplateSelectorChanged(DataTemplateSelector
      oldContentTemplateSelector, DataTemplateSelector newContentTemplateSelector);
    protected virtual void OnTemplateChanged(DataTemplate oldTemplate,
            DataTemplate newTemplate);
    public bool ShouldSerializeContentTemplateSelector();
}
```

分析 ContentPresenter 对象模型，理解其工作原理。首先，ContentPresenter 检查内容包含的数据类型，若是 System.Windows.UIElement（控件的基类型），把内容直接添加到可视化树中即可；若 ContentTemplate 有值，则创建一个 UIElement 实例，并把实例添加到可视化树中；若 ContentTemplateSelector 有值，通过它找到模板，并使用模板创建一个 UIElement 实例，并把实例添加到可视化树中；若是能转换到字符串的 TypeConverter 实例，可以把 Content 封装到 TextBlock 里，并把这个 TextBlock 添加到可视化树。最后，调用 Content 的 ToString 方法，把返回结果封装到 TextBlock 中，把 TextBlock 添加到可视化树中。

上例是 Button 按钮通过内容呈现器实现简单的编程模型，体现了元素合成的思想。对于内容模型，有 Content、Items、Child、Children 4 种通用模式，并通过内容属性命名 4 种模式，如表 4.1 所示。

表 4.1　内容属性命名模式

单/复数	对象	元素
单一	Content	Child
多重	Items	Children

了解 ContentPresenter 如何呈现数据后，需要再学习 Items 对象、Child 和 Children 两元素。

2. Items

ContentPresenter 适用于单一的内容模式，若是多重内容模式，则需用 Items 属性，CS 代码如下。

```
public MainWindow()
    {
        InitializeComponent();
        ListBox l = new ListBox();
        l.Items.Add("Hello");
        l.Items.Add("WPF");
    }
```

3. Child 和 Children

Button 和 Listbox 的内容可以是任意对象,而有些控件内容只能是 UIElement 类型,其对应的内容模式就是 Child 和 Children。

在深入学习 Child 和 Children 内容模式之前,先来了解一下 WPF 控件的类型。WPF 的架构师把 WPF 控件大致可以分为三类:内容控件(Content Controls)、布局控件(Layout Controls)和呈现控件(Render Controls)。

前面提到的 Button 和 ListBox 属于内容控件,它们需要与其他元素一起工作,由图 4.1 的 Button 可视化树显示,ButtonChrome、ContentPresenter 和 TextBlock 3 个元素合成一个 Button。其中,布局控件定位其他控件,支持多重内容模式。除了 FlowDocumentViewer 以外的所有布局面板,都继承自 Panel。多重元素内容模型如下。

```
public abstract class Panel : FrameworkElement, IAddChild
{
    protected Panel();
    public UIElementCollection Children { get; }
}
```

呈现控件就是把内容显示在屏幕上的绘制控件。Ellipse 和 Rectangle 是最常见的呈现控件。Border 类是为单一元素添加边框的呈现控件,内容模型如下。

```
public class Border : Decorator
{
    public Brush Background { get; set; }
    public Brush BorderBrush { get; set; }
    public Thickness BorderThickness { get; set; }
    public CornerRadius CornerRadius { get; set; }
    public Thickness Padding { get; set; }
}
```

内容模型解决了编程控件内容的多样性问题。控件以何种方式呈现其外观,即控件对外的表现形式又是怎样体现呢?在 Button 按钮内部的 ButtonChrome 是如何呈现的呢?下面通过学习模板来理出头绪。

4.1.2 模板

WPF 控件具有模板系统,通过属性来对改变其外观。ButtonChrome 是由按钮模板创建出来的,是 Button 的内容呈现器。

在先前内容模型中,将 WPF 控件大致分为内容、布局和呈现控件 3 种类型。其中,所有的内容控件都支持模板,模板可以创建一种特别的元素(如 ButtonChrome)来显示其外观。内容控件继承自 Control，并具有通用名为 Template 的属性。

在 XAML 文件中，编写一个 Button 定义代码：<Button>The Button</Button>。在不影响该按钮内容的前提下，若要改变这个按钮的外观，需要定义一个新的模板。模板可以看作可视化树的创建工厂。这就是说，创建模板后，ControlTemplate 为控件创建可视化树；DataTemplate 为数据创建可视化树。

定义模板时，要明确模板的目标类型和可视化树。XAML 代码如下。

```
<Button>
    <Button.Template>
        <ControlTemplate TargetType = "{x:Type Button}">
            <Rectangle Fill = "Green" Width = "80" Height = "25"></Rectangle>
        </ControlTemplate>
    </Button.Template>
    The Button
</Button>
```

运行上述代码，生成一个绿色矩形 Button，如图 4.5 所示。其对应的可视化树如图 4.6 所示。该可视化树是单一的绿色矩形。

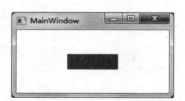

图 4.5 以绿色矩形作模板的按钮

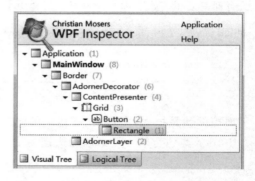

图 4.6 绿色矩形作模板的按钮的可视化树

观察绿色矩形作模板的按钮的可视化树，只有一个绿色的 Rectangle。在按钮的可视化树中的 ButtonChrome、ContentPresenter 和 TextBlock 在模板中是不存在的。接下来为 Button 添加 Click 事件。Click() 事件实现的功能是：创建 Button 的新模板，让 Button 变成椭圆形，CS 代码如下。

```
private void Button_Click(object sender, RoutedEventArgs e)
{
    ControlTemplate template = new ControlTemplate(typeof(Button));
    template.VisualTree = new FrameworkElementFactory(typeof(Ellipse));
    template.VisualTree.SetValue(Ellipse.FillProperty, Brushes.Purple);
    template.VisualTree.SetValue(Ellipse.WidthProperty, 100.0);
    template.VisualTree.SetValue(Ellipse.HeightProperty, 30.0);
    ((Button)sender).Template = template;
}
```

把 Click() 事件与 Button 建立关联，在 XAML 代码中补充代码如下。

```
< Button Click = "Button_Click">
…
</Button>
```

单击绿色矩形按钮,页面显示效果如图 4.7 所示。其对应的可视化树如图 4.8 所示。该可视化树是单一的紫色椭圆矩形。

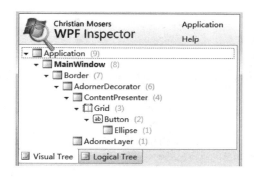

图 4.7　以紫色椭圆模板的按钮　　　　图 4.8　紫色椭圆模板的按钮的可视化树

1. 模板绑定

由前面的学习中,可以得出 WPF 控件的 3 个原则:元素合成、内容的多样性和简单的编程模型。其中,元素合成和内容的多样性两个原则是可以通过模板来实现它们。然而,改变按钮的外观通过定义新的模板来实现,这并不是简单的编程模型。

在此,使用模板的属性,在模板控件上绑定属性,让用户控件上的属性来自定义模板。XAML 代码如下。

```
<Grid>
    <Button
        BorderThickness = "3"
        BorderBrush = "GreenYellow"
        Background = "Bisque"
        Width = "100"
        Height = "30">
        <Button.Template>
            <ControlTemplate TargetType = "{x:Type Button}">
                <Border CornerRadius = "6"
                BorderThickness = "{TemplateBinding Property = BorderThickness}"
                BorderBrush = "{TemplateBinding Property = BorderBrush}"
                Background = "{TemplateBinding Property = Background}">
                    <ContentPresenter/>
                </Border>
            </ControlTemplate>
        </Button.Template>
    </Button>
</Grid>
```

运行上述代码,页面显示效果如图 4.9 所示,模板绑定按钮的可视化树如图 4.10 所示。XAML 把边框(Border)的 BorderThickness、BorderBrush 和 Background 属性绑到了 Button 模板。注意,绑定时属性要一致。实现如图 4.9 所示的页面效果还可以用下面的 XAML 代码。

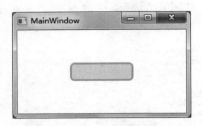

图 4.9 模板绑定按钮

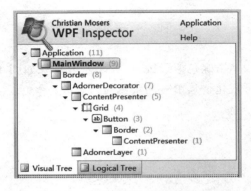

图 4.10 模板绑定按钮的可视化树

```
<!-- 保留与上例相同代码部分 -->
<Button.Template>
        <ControlTemplate TargetType = "{x:Type Button}">
            <Border CornerRadius = "6"
            BorderThickness = "{TemplateBinding Property = BorderThickness}"
            BorderBrush = "{TemplateBinding Property = BorderBrush}">
                <Rectangle Fill = "{TemplateBinding Property = Background}"/>
            </Border>
        </ControlTemplate>
</Button.Template>
<!-- 保留与上例相同代码部分 -->
```

运行上述代码,页面显示效果与图 4.9 完全相同,但是它们的可视化树是不一样的,请读者仔细分析和辨别。

2. 模板思维

WPF 新理念中还隐含着增量式的自定义控件思想。使用 WPF 的模板,用户界面的任意部分都能自定义,可以创建更友好的交互模型。Windows 属于内容控件,现在为它做一个模板,XAML 代码如下。

```
<!-- 保留 Window 代码部分 -->
    Title = "Custom Window" Height = "350" Width = "525">
    <Window.Template>
        <ControlTemplate TargetType = "{x:Type Window}">
            <Border BorderThickness = "2" BorderBrush = "Purple"
            Background = "GreenYellow">
              <Grid>
                    <Rectangle Fill = "Pink" Margin = "8" />
                    <TextBlock Margin = "90" Text = "{TemplateBinding Title}"/>
                </Grid>
            </Border>
        </ControlTemplate>
    </Window.Template>
</Window>
```

运行上面的代码,页面显示效果如图 4.11 所示,窗体模板的可视化树如图 4.12 所示。在今后应用程序的设计中,开发人员可采用模板技术,为应用程序做出个性化的页面,使用模板思考,设计出更友好的交互页面。

图 4.11　窗体模板

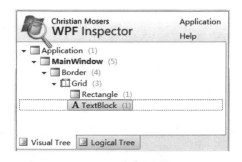

图 4.12　窗体模板的可视化树

在了解过 WPF 控件中的内容模型与模板的新理念以后,再来学习 WPF 中常用的控件。在上一节的中将 WPF 控件大致分为内容、布局和呈现控件 3 种类型。其中布局控件在第 3 章中已详细论述,这里不再赘述。接下来,对内容和呈现两类控件要做更细致的划分。按照控件在应用程序的主页面上出现的先后顺序,先来学习 WPF 中的菜单、工具栏和状态栏控件。

4.2　菜单、工具栏和状态栏

菜单与工具栏位于主页面的顶部,两者实现的功能是相同的。不同之处是：显示位置和交互模型。通常情况下,菜单有层级关系,并固定占用了顶部很少的屏幕；工具栏在菜单下方,它是菜单中的某些命令或快捷方式。菜单和工具栏共同实现程序功能,方便用户操作。

4.2.1　Menu

菜单在基于元素合成的原则下,包含托管在 Menu 或 ContentMenu 中的 MenuItem 控件。在 Windows 应用程序中,其位于窗口的顶部。WPF 菜单可分为 Menu 和 ContextMenu 两种。其中,ContextMenu 称为上下文菜单,只有当用户发出请求(右击)或按下 Shift+F10 组合键时,会弹出上下文菜单,故上下文菜单又可称为弹出式菜单。

通常情况下,菜单位于窗口的顶部,但在 WPF 中,菜单可以放到任意位置。

因菜单具有层级关系,在创建菜单时,需要添加具有层级关系的 MenuItem 控件到 Menu 对象中,并且用 DockPanel 布局控件,才能让菜单显示在窗口的顶部。接下来创建菜单,XAML 代码如下：

```
<Window x:Class = "Menu.MainWindow"
        xmlns = "http://schemas.microsoft.com/winfx/2006/xaml/presentation"
        xmlns:x = "http://schemas.microsoft.com/winfx/2006/xaml"
        Title = "MainWindow" Height = "350" Width = "525">
    <DockPanel LastChildFill = "False">
        <Menu DockPanel.Dock = "Top">
            <MenuItem Header = "_File">
                <MenuItem Header = "_Open" Click = "Open_Clicked"/>
                <MenuItem Header = "_Exit" Click = "Exit_Clicked"/>
```

```
            </MenuItem>
            <MenuItem Header="_Edit">
                <MenuItem Header="_Cut"/>
                <MenuItem Header="_Copy"/>
            </MenuItem>
        </Menu>
    </DockPanel>
</Window>
```

运行上述代码,单击 File 菜单,页面显示效果如图 4.13 所示。单击 Edit 菜单,页面显示效果如图 4.14 所示。代码中的"LastChildFill="False""表示最后一个子元素不拉伸。

图 4.13　File 菜单窗口页面

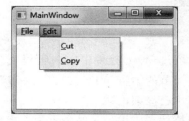

图 4.14　Edit 菜单窗口页面

现在的菜单还不具有交互功能,接下来对 File 菜单下的 Open 与 Exit 菜单项编写 Click 事件,CS 代码如下。

```
public partial class MainWindow : Window
{
    public MainWindow()
    {
        InitializeComponent();
    }
    void Open_Clicked(object sender, RoutedEventArgs e)
    {
        OpenFileDialog ofd = new OpenFileDialog();
        ofd.Filter = "文本文件|*.txt|图片|*.jpg|所有文件|*.*";  //过滤器
        if (ofd.ShowDialog() == true)
        {
            string file_name = ofd.FileName;            //获取打开文件的路径
        }
    }
    void Exit_Clicked(object sender, RoutedEventArgs e)
    {
        Close();
    }
}
```

运行上述代码后,选择 File 菜单下的 Open 菜单项,打开对话框;选择 File 菜单下的 Exit 菜单项,结束程序。

4.2.2　ToolBar

工具栏位于菜单下方,具有宿主类型(ToolBarTray)和项类型(ToolBar)。ToolBarTray

可视为 ToolBar 的容器，用于调整工具栏的大小及位置，续上例，在菜单窗口，加入工具栏的 XAML 代码如下。ToolBar 通过 Band 和 BandIndex 两属性来定位。

```
<!-- 保留与上例相同代码部分 -->
    <DockPanel LastChildFill = "False">
        <Menu>
            ...
        </Menu>
        <ToolBarTray DockPanel.Dock = "top">
            <ToolBar>
                <Button>Open</Button>
                <Button>Cut</Button>
                <Button>Copy</Button>
                <Button>Paste</Button>
                <Button>Exit</Button>
            </ToolBar>
            <ToolBar Header = "Search">
                <TextBox Width = "120"/>
                <Button Width = "25">Go</Button>
            </ToolBar>
        </ToolBarTray>
    </DockPanel>
</Window>
```

运行上述代码，页面显示效果如图 4.15 所示。将含有 Search 按钮的工具栏移动到 File 菜单正下方，如图 4.16 所示。

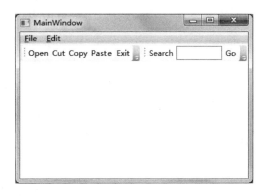

图 4.15　菜单和工具栏窗口

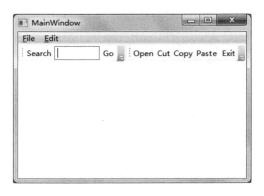

图 4.16　被移动后的菜单和工具栏窗口

工具栏可用做控件容器，如本例中的搜索框就是一个文本框，生活中常用的 Web 浏览器中的地址栏则是一个组合框。

4.2.3　StatusBar

状态栏控件通常位于窗体底部，用于显示状态文本信息。StatusBar 也是容器控件，要呈现信息的控件是它的子元素。续上例，在菜单工具栏窗口加入状态栏的 XAML 代码如下。

```
<!-- 保留与上例相同代码部分 -->
        </ToolBarTray>
            <StatusBar DockPanel.Dock = "Bottom" Background = "LightBlue">
                <TextBox Text = "This is a StatusBar"></TextBox>
            </StatusBar>
        </DockPanel>
</Window>
```

运行上述代码,页面显示效果如图 4.17 所示。本例中的 TextBox 是 StatusBar 的子元素,用来显示状态信息。

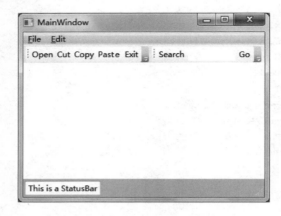

图 4.17 菜单、工具和状态栏窗口

在应用程序中,状态信息是随着程序的运行动态变化的,故此其内部元素应该命名,便于后台操作,同时应该设置窗体的属性,让状态栏的右下角呈现调整区域,设置窗体的属性:ResizeMode="CanResizeWithGrip"。

4.3 容器控件

容器控件能容纳子控件,并为子控件提供可视化的分组功能。在第 3 章介绍的布局控件都可视为容器控件。本节重点讲述 Expander、GroupBox 和 TabControl 这 3 个控件。

4.3.1 Expander

Expander 是可以展开和折叠的控件,由标题头和内容两部分组成。Header 属性设置标题头;Content 属性设置内容。下面构建 Expander 控件,XAML 代码如下。

```
<Grid>
    <Expander Header = "Expander example">
        <Border Margin = "6" Padding = "6">
            <StackPanel>
                <TextBox>I am first textbox</TextBox>
                <TextBox>I am second textbox</TextBox>
            </StackPanel>
        </Border>
```

```
        </Expander>
</Grid>
```

运行上述代码,页面显示效果如图 4.18 所示,默认折叠状态,内容隐藏,但是占有空间。单击标题头,展开后的页面显示效果如图 4.19 所示。如果让代码在运行后,即为展开状态,则要设置属性"IsExpanded="True""。

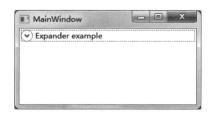

图 4.18　Expander 默认折叠效果

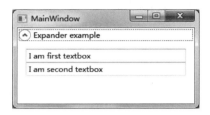

图 4.19　Expander 展开效果

4.3.2　GroupBox

GroupBox 是对控件进行分组的可视化容器控件,并将同类控件归类,置于 GroupBox 控件中。接下来,续上例,在 Expander 的基础上加入 GroupBox 控件的 XAML 代码如下。

```
<!-- 保留与上例相同代码部分 -->
        </Expander>
        <GroupBox Header = "GroupBox example" Margin = "15,100,15,15">
            <StackPanel>
                <RadioButton Content = "Male" IsChecked = "True"/>
                <RadioButton Content = "Female"/>
            </StackPanel>
        </GroupBox>
</Grid>
```

运行上述代码,页面显示效果如图 4.20 所示,GroupBox 位于 Expander 下方,两者是并列关系,案例中的控件有嵌套关系,使用逻辑树来表示多控件层级嵌套,如图 4.21 所示。

图 4.20　GroupBox 分组页面

图 4.21　多控件层级嵌套逻辑树

4.3.3 TabControl

TabControl 支持传统的标签风格，设计用户界面时，常会用选项卡设置、标签式浏览器等。接下来，在上例的基础上加入 TabControl 控件，XAML 代码如下。

```
<Grid>
    <TabControl>
        <TabItem Header = "Tab1">
            <StackPanel>
                <!-- 保留与上例相同代码部分 -->
            </StackPanel>
        </TabItem>
        <TabItem Header = "Tab2"/>
    </TabControl>
</Grid>
```

运行上述代码，页面显示效果如图 4.22 所示。该例中的控件有多重嵌套关系，逻辑树表示多重层级嵌套关系，如图 4.23 所示。

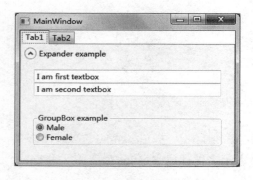

图 4.22 TabControl 控件页面

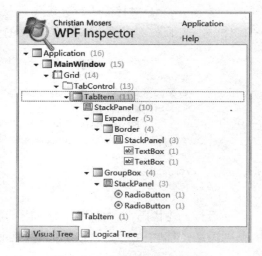

图 4.23 多重层级嵌套逻辑树

本节中的 3 个控件都继承自 HeaderedContentControl，HeaderedContentControl 又继承自 ContentControl。故此，有些书上也把它们称为带标题的内容控件。

4.4 范围控件

WPF 具有 Slider、ScrollBar 和 ProgressBar 3 个范围控件。这些控件在规定的范围内取值，继承自 RangeBase，RangeBase 又继承自 Control。在此说明 RangeBase 类的公共属性，如表 4.2 所示。

表 4.2　RangeBase 的公共属性

属 性 名	属 性 描 述
Maximum	上限的最大值
Minimum	下限的最小值
LargeChange	属性值的最大变化
SmallChange	属性值的最小变化
Value	控件当前值

当 Value 值改变时，Slider、ScrollBar 事件响应，但是 ProgessBar 控件在 Value 变化时不响应，它能响应 Click、Dragdrop、Dragover、Mousemove、Mouseup 和 Mousedown 这 6 个事件。

4.4.1　Slider

Slider 控件允许用户在可视化的最小值和最大值范围之间取值，更直观。下面学习 Slider 控件，XAML 代码如下。

```
< Slider Minimum = "0" Maximum = "100" Value = "50" IsSnapToTickEnabled = "True"
        TickPlacement = "Both" TickFrequency = "5">
</Slider >
```

运行上述代码，页面显示效果如图 4.24 所示，带有记号的刻度提升用户体验。

在设计时，时常会遇到 Slider 受限在一个较小的范围内，此时可用 3 个属性来设置，XAML 代码如下。

```
< Slider Minimum = "0" Maximum = "100" Value = "50" IsSnapToTickEnabled = "True"
        TickPlacement = "BottomRight" TickFrequency = "5" IsSelectionRangeEnabled = "True"
        SelectionStart = "50" SelectionEnd = "90">
</Slider >
```

运行上述代码，页面显示效果如图 4.25 所示，带有范围选择的 Slider。Slider 控件中用到了几个属性，它们代表的含义如表 4.3 所示。

图 4.24　带刻度的 Slider

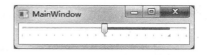
图 4.25　带范围选择的 Slider

表 4.3　Slider 的属性

属 性 名	属 性 描 述
IsSnapToTickEnabled	是否加标记
TickPlacement	标记位置
TickFrequency	标记间隔大小
IsSelectionRangeEnabled	是否加范围选择
SelectionStart	选择范围初始值
SelectionEnd	选择范围终止值

4.4.2 ScrollBar

ScrollBar 控件是滚动条状态，通过设置"Orientation＝"Horizontal""，为水平滚动。创建 ScrollBar 控件的 XAML 代码如下。

```
< ScrollBar Orientation = "Horizontal" Width = "250" Height = "20"
            Margin = "10" Background = "LightSalmon">
</ScrollBar >
```

运行上述代码，页面显示效果如图 4.26 所示。再给出 ScrollBar 属性的取值，XAML 代码如下。

```
< ScrollBar Orientation = "Horizontal" Margin = "10" Width = "250" Height = "20"
            Background = "LightSalmon" Minimum = "8" Maximum = "340" Value = "98">
</ScrollBar >
```

运行上述代码，页面显示效果如图 4.27 所示。

图 4.26　ScrollBar

图 4.27　带取值范围的 ScrollBar

4.4.3 ProgressBar

ProgressBar 称为进度条，指示当前任务的工作状态，不与用户发生交互，也不响应鼠标事件和键盘输入。在大多数情况下，无法获知任务执行时间，将 IsIndeterminate 属性设为 True，此时，在窗口中添加 ProgressBar 控件，XAML 代码如下。

```
< ProgressBar Width = "200" Height = "15" IsIndeterminate = "True" />
```

运行上述代码，页面显示效果如图 4.28 所示，ProgressBar 周期性地显示一个从左向右的绿色进度条，这样来表示任务正在执行。在代码中加入背景色及前景色"Background＝"Orange" Foreground＝"LawnGreen""属性设置后，页面效果如图 4.29 所示。进度条可以用在连接远程服务器、程序安装过程、下载数据等情况，给用户一个系统状态反馈。

图 4.28　系统默认 ProgressBar 的样式

图 4.29　改变背景的 ProgressBar

4.5　文本编辑器控件

WPF 提供的文本编辑器控件有 PasswordBox、TextBox、RichTextBox 和 InkCanvas。本节首先需了解文本对象模型。

4.5.1 文本模型

WPF 中处理文本数据有集合模型和流文本模型两种模型。集合模型适用于动态文本构建和文本校验。流文本模型则适用于富文本进行编辑。

集合模型类似控件的属性用法，构造文本元素后，将其添加到其他文本元素的集合上。动态添加文本的 CS 代码如下。

```
private void Button_Click(object sender, RoutedEventArgs e)
{
    FlowDocument document = new FlowDocument();
    Paragraph para = new Paragraph();
    para.Inlines.Add(new Run("Hello WPF"));
    document.Blocks.Add(para );
}
```

文本中的两个主要元素是块元素和行元素。块元素的载体是矩形框，块元素的示例是 Paragraph 或 Table，行元素横跨多个行，行元素的示例是 Span、Run 和 Bold。块元素对应着 FlowDocument 中的 Blocks 属性；行元素对应着 Paragraph 的 Inlines 属性。

流文本模型把文本当作一个流来操作，不太容易理解。人们所熟知的 UI 函数库（User32、Windows Forms、Abstract Window Toolkit 等）把元素表示为树，每个元素有一个父对象和一些子对象。

在典型的文本函数库（如 Internet Explorer）把其中的对象表示为一个文本流。因为函数库构建部件是一个字符串流，所以其中的元素都视为一个文本流。下面将学习文本编辑器控件。

4.5.2 PasswordBox

PasswordBox 的功能与文本框类似，但它会把用户的输入显示为圆点或星形，它并不支持文本模型，其独特的功能是存储用户密码。

4.5.3 TextBox 与 RichTextBox

TextBox 与 RichTextBox 用法相似，但是 TextBox 不支持富文本，是对 RichTextBox 的编辑功能的简化。创建 TextBox 的 XAML 代码如下。

```
< TextBox AcceptsReturn = "True" TextWrapping = "Wrap" FontSize = "34"
    Text = "Hello TextBox"/>
```

运行上述代码，页面显示效果如图 4.30 所示。其中，"TextWrapping＝"Wrap""实现换行输入功能，"AcceptsReturn＝"True""实现换行输出功能。

RichTextBox 是功能最强大的文本编辑器，支持大约 84 个命令，创建 RichTextBox 的 XAML 代码如下。

```
<!-- 保留 Window 代码部分 -->
```

```
            <RichTextBox>
                <FlowDocument FontSize = "34">
                    <Paragraph>Hello RichTextBox</Paragraph>
                </FlowDocument>
            </RichTextBox>
        </Window>
```

运行上述代码，页面显示效果如图 4.31 所示。RichTextBox 是支持流文本模型，是 FlowDocument 编辑器。它支持常见的编辑命令及常用的快捷键（如 Ctrl＋A、Ctrl＋C、Ctrl＋Z 等）。

图 4.30　TextBox 实现文本换行输入

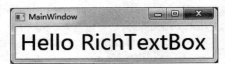

图 4.31　RichTextBox 编辑器

4.5.4　InkCanvas

InkCanvas 是针对手写笔迹的编辑器。笔迹数据比富文本简单，可视做一个数字化接收器。数字化接收器的功能是把手写的模拟信号转化为数字信号。笔迹基础数据是 Stroke，它定义在 System.Windows.Ink 名称空间中。为了解笔迹的工作机制，在 InkCanvas 中的 XAML 代码如下。

```
        <!-- 保留 Window 代码部分 -->
        <Grid>
            <InkCanvas
                StrokeCollected = "InkProcess"
                Background = "Beige"/>
            <Canvas Name = "layover"></Canvas>
        </Grid>
    </Window>
```

上述代码中，InkCanvas 控件的 InkProcess 事件的后台 CS 代码如下。

```
private void InkProcess(object sender, InkCanvasStrokeCollectedEventArgs e)
{
    layover.Children.Clear();
    Brush fill = new SolidColorBrush(Color.FromArgb(150, 200, 0, 0));
    foreach (StylusPoint pt in e.Stroke.StylusPoints)
    {
        double markerSize = pt.PressureFactor * 35;
        Ellipse marker = new Ellipse();
        Canvas.SetLeft(marker, pt.X - markerSize/2);
        Canvas.SetLeft(marker, pt.Y - markerSize/2);
        marker.Width = marker.Height = markerSize;
        marker.Fill = fill;
```

```
            layover.Children.Add(marker);
        }
}
```

运行完整的 InkCanvas 代码,接收手写数据后,页面显示效果如图 4.32 所示。对于 InkCanvas 还支持笔势,作为笔势识别器可以分析笔迹数据。启用笔势有两个步骤。第一步,设置 InkCanvas 的 EditingMode 属性值为 InkAndGesture;第二步,通知笔势识别器查找笔势时,需要调用 InkCanvas 上的 SetEnabledGestures。XAML 代码如下。

图 4.32　InkCanvas 工作对象模型

```
<StackPanel>
    <InkCanvas Height = "150" Name = "_ink"
            Gesture = "InkGesture" EditingMode = "InkAndGesture">
    </InkCanvas>
    <ListBox Name = "_look"></ListBox>
</StackPanel>
```

InkCanvas 控件后台的 CS 代码如下。

```
public MainWindow()
{
    InitializeComponent();
    _ink.SetEnabledGestures(new ApplicationGesture[]{
        ApplicationGesture.AllGestures,});
}
private void InkGesture(object sender, InkCanvasGestureEventArgs e)
{
    _look.Items.Add(e.GetGestureRecognitionResults()[0].ApplicationGesture);
}
```

运行这个完整的 InkCanvas 的手势识别程序,写下笔势,如图 4.33 所示。

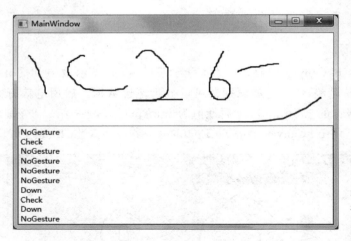

图 4.33 InkCanvas 笔势识别器

4.6 列表控件

在应用程序中,常见到显示数据的列表。WPF 中列表控件的主要功能是显示数据和选择数据。WPF 中有 4 个标准列表控件,即 ListBox、ComboBox、ListView 和 TreeView。列表控件通过 Items 和 ItemsSource 两属性添加数据源。Items 逐条添加数据源到列表控件中,代码如下。

```
ListBox list = new ListBox();
list.Items.Add("one");
list.Items.Add("two");
```

ItemsSource 是以集合的方式添加数据源到列表控件中,代码如下。

```
string[] items = new string[] {"one","two"};
ListBox list1 = new ListBox();
list1.ItemsSource = items;
```

对比添加数据源的两种方法,使用 ItemsSource 可在列表控件外部维护数据。

4.6.1 ListBox 和 ComboBox

尽管 ListBox 和 ComboBox 的显示页面差别很大,但从对象模型的角度看,两者又很相似。故把两个控件放到一起,对比学习。

由 WPF 控件内容模型和模板的新理念可知,可以把任意类型的数据放到列表控件中,并用模板来改变控件的外观。

为了更好地理解 ListBox 的应用场合,选取 Windows 7 控制面板如图 4.34 所示。

图 4.34 中的控制面板就是使用 Grid 布局的 ListBox 设计而成的。下面用自定义列表的控件 ItemsPanel 创建新的布局模板,用于显示列表中的项目。XAML 代码如下。

```
<!-- 保留 Window 代码部分 -->
```

图 4.34 Windows 7 控制面板

```
<Grid>
    <ListBox>
        <ListBox.ItemsPanel>
            <ItemsPanelTemplate>
                <UniformGrid Columns = "4"></UniformGrid>
            </ItemsPanelTemplate>
        </ListBox.ItemsPanel>
        <ListBoxItem> Item1 </ListBoxItem>
        <ListBoxItem> Item2 </ListBoxItem>
        <ListBoxItem> Item3 </ListBoxItem>
        <ListBoxItem> Item4 </ListBoxItem>
        <ListBoxItem> Item5 </ListBoxItem>
        <ListBoxItem> Item6 </ListBoxItem>
        <ListBoxItem> Item7 </ListBoxItem>
        <ListBoxItem> Item8 </ListBoxItem>
        <ListBoxItem> Item9 </ListBoxItem>
        <ListBoxItem> Item10 </ListBoxItem>
        <ListBoxItem> Item11 </ListBoxItem>
        <ListBoxItem> Item12 </ListBoxItem>
    </ListBox>
</Grid>
</Window>
```

运行上述代码，页面显示效果如图 4.35 所示。使用 UniformGrid 布局列表框，Columns 属性设置列表的列数。

ComboBox 从用户的角度来看，就是由一个文本输入控件和一个下拉菜单组成的，用户可以从一个预先定义的列表中选择一个选项，也可以直接在文本框中输入文本。它占据的屏幕空间比 ListBox 更小，而且数据项可以隐藏。创建 ComboBox 的 XAML 代码如下。

```
<! -- 保留 Window 代码部分 -->
<StackPanel Margin = "10">
    <ComboBox>
        <ComboBoxItem> ComboBox Item1 </ComboBoxItem>
        <ComboBoxItem IsSelected = "True"> ComboBox Item2 </ComboBoxItem>
        <ComboBoxItem> ComboBox Item3 </ComboBoxItem>
        <ComboBoxItem> ComboBox Item4 </ComboBoxItem>
        <ComboBoxItem> ComboBox Item5 </ComboBoxItem>
    </ComboBox>
</StackPanel>
</Window>
```

运行上述代码，页面显示效果如图 4.36 所示。"IsSelected = "True""所在的位置表示当前 ComboBox 的内容是程序运行后的默认项。

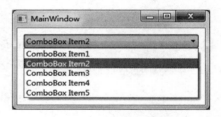

图 4.35　ListBox 控件　　　　　图 4.36　ComboBox 控件

4.6.2　ListView

ListView 在 ListBox 的基础上又有新的特性，它可以看作可视化的网格控件，所以在使用 ListView 时，要把 ListView.View 属性设置为 GridView。使用 ListView 显示含有表头为姓名、性别和房间名的表格。XAML 代码如下。

```
<!-- 保留 Window 代码部分 -->
Title = "大观园" Height = "350" Width = "525">
    < Grid >
        < ListView Name = "lv" ItemsSource = "{Binding}">
            < ListView.View >
                < GridView AllowsColumnReorder = "True">
                    < GridViewColumn Width = "50" Header = "姓  名"
                        DisplayMemberBinding = "{Binding Path = Name}" />
                    < GridViewColumn Width = "50" Header = "性  别"
                        DisplayMemberBinding = "{Binding Path = Gender}"/>
                    < GridViewColumn Width = "55" Header = "房间名"
                        DisplayMemberBinding = "{Binding Path = RoomName}" />
                </GridView >
            </ListView.View >
        </ListView >
    </Grid >
</Window >
```

在后台添加数据项的 CS 代码如下。

```
using System.Collections.ObjectModel;        //新增添引用
public ObservableCollection < object > ObservableObj;
public MainWindow()
{
    InitializeComponent();
    ObservableObj = new ObservableCollection < object >();
    ObservableObj.Add(new { Name = "贾宝玉", Gender = "男", RoomName = "怡红院"});
    ObservableObj.Add(new { Name = "林黛玉", Gender = "女", RoomName = "潇湘馆"});
    ObservableObj.Add(new { Name = "薛宝钗", Gender = "女", RoomName = "蘅芜院"});
    ObservableObj.Add(new { Name = "贾迎春", Gender = "女", RoomName = "紫菱洲"});
    ObservableObj.Add(new { Name = "贾探春", Gender = "女", RoomName = "秋爽斋"});
    ObservableObj.Add(new { Name = "贾惜春", Gender = "女", RoomName = "暖香坞"});
    ObservableObj.Add(new { Name = "妙玉",   Gender = "女", RoomName = "拢翠庵"});
```

```
        ObservableObj.Add(new { Name = "李纨", Gender = "女", RoomName = "稻香村" });
        lv.DataContext = ObservableObj;
}
```

运行完整的代码,页面显示效果如图4.37所示。CS代码中使用Add()方法逐条添加数据,较为烦琐。通常数据与ListView绑定是常用的方法,其XAML代码如下。

```
<!-- 保留Window代码部分 -->
Title = "ListViewDataBindingExample" Height = "350" Width = "525">
    <Grid>
        <ListView ItemsSource = "{x:Static Fonts.SystemFontFamilies}" Margin = "0,0,0,-12">
            <ListView.View>
                <GridView>
                    <GridViewColumn Header = "Name"
                            DisplayMemberBinding = "{Binding Source}"/>
                    <GridViewColumn Header = "Line Spacing"
                            DisplayMemberBinding = "{Binding LineSpacing}"/>
                    <GridViewColumn Header = "Demo">
                        <GridViewColumn.CellTemplate>
                            <DataTemplate>
                                <TextBlock FontFamily = "{Binding}" FontSize = "16"
                                        Text = "I am a GridViewColumn"/>
                            </DataTemplate>
                        </GridViewColumn.CellTemplate>
                    </GridViewColumn>
                </GridView>
            </ListView.View>
        </ListView>
    </Grid>
</Window>
```

运行上述代码,页面显示效果如图4.38所示。其中,列表中的Name与Line Spacing使用的是绑定的系统值,Demo使用的是数据模板。

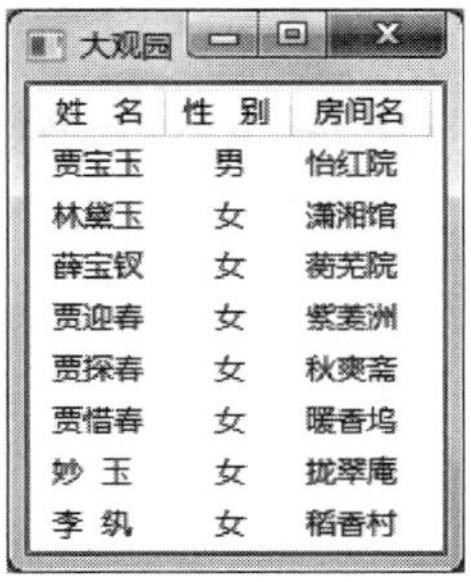

图4.37 ListView控件

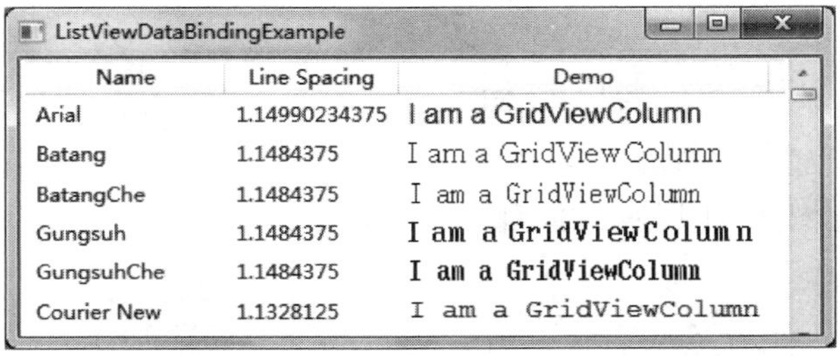

图4.38 具有数据绑定的ListView控件

4.6.3 TreeView

TreeView可以视为列表框的嵌套,也就是说列表框中的条目上还是一个列表框。TreeView把每个TreeView对象看作一个控件。XAML创建TreeView的代码如下。

```xml
<!-- 保留 Window 代码部分 -->
    <Grid>
        <TreeView>
            <TreeViewItem Header = "Name">
                <TreeViewItem Header = "Tom"/>
                <TreeViewItem Header = "Alice"/>
            </TreeViewItem>
            <TreeViewItem Header = "Hobby">
                <TreeViewItem Header = "Sing"/>
                <TreeViewItem Header = "Draw"/>
            </TreeViewItem>
        </TreeView>
    </Grid>
</Window>
```

运行上述代码,页面显示效果如图 4.39 所示。和其他列表控件使用相似,直接添加条目到 Items 属性上。

给 TreeView 添加条目,通过 Add()方法实现的 CS 代码如下。

```
public MainWindow()
{
    InitializeComponent();
    TreeViewItem item1 = new TreeViewItem() { Header = "北京" };
    TreeViewItem item11 = new TreeViewItem() { Header = "故宫" };
    item11.Items.Add("南大门");
    item11.Items.Add("神武门");
    item11.Items.Add("东华门");
    item11.Items.Add("西华门");
    item1.Items.Add(item11);
    item1.Items.Add("颐和园");
    item1.Items.Add("水立方");
    tw.Items.Add(item1);
}
```

在 XAML 代码中,写入< TreeView Name = "tw"></TreeView >语句后再运行代码,页面显示效果如图 4.40 所示。请读者在图 4.39 页面的基础上,添加 Checkbox 控件后,做出如图 4.41 所示的页面。

图 4.39　TreeView 控件　　图 4.40　CS 编写 TreeViewItem　　图 4.41　带 Checkbox 的 TreeView

4.7 构 建 控 件

WPF 的控件模型中采用元素合成的思想，可以用较小的控件构建较大的控件，在此来了解系统为构建控件所提供的服务，哪些服务适用于构建自定义控件。其中构建控件的较小组件都在 System.Windows.Controls.Primitives 名称空间下。

4.7.1　ToolTip

在 WPF 中，ToolTip 既可以表示工具提示属性，还可以表示一个类。当它作为 Button 控件属性时，在 XAML 代码中，设置 ToolTip 的代码如下。

```xml
<Button ToolTip = "please click to enter next step"></Button>
```

在 ToolTip 类中，可以添加任何的控件来修饰这个控件的 ToolTip，代码如下。

```xml
<!-- 保留 Window 代码部分 -->
<Grid>
    <Button Height = "25" Content = "工具提示演示" HorizontalAlignment = "Left">
        <Button.ToolTip>
            <ToolTip Background = "Green" Foreground = "White" HasDropShadow = "False"
                     Placement = "MousePoint">
                <TextBlock Margin = "5">ToolTip 类的使用方法</TextBlock>
            </ToolTip>
        </Button.ToolTip>
    </Button>
</Grid>
</Window>
```

运行上述代码，页面显示效果如图 4.42 所示。其中，ToolTip 使用 TextBlock 作为工具提示控件。还可以使用 ToolTipService 显示工具提示信息，XAML 代码如下。

```xml
<!-- 保留 Window 代码部分 -->
Title = "MainWindow" Height = "350" Width = "525"
ToolTipService.InitialShowDelay = "0" ToolTipService.ShowDuration = "99999">
<Window.ToolTip>
    <ToolTip x:Name = "_tooltip" Placement = "RelativePoint" VerticalOffset = "15">
    </ToolTip>
</Window.ToolTip>
    <UniformGrid Rows = "2" Columns = "3">
        <Button Margin = "6">one</Button>
        <Button Margin = "6">two</Button>
        <Button Margin = "6">three</Button>
        <Button Margin = "6">four</Button>
        <Button Margin = "6">five</Button>
        <Button Margin = "6">six</Button>
    </UniformGrid>
</Window>
```

后台 CS 代码如下。

```
public partial class MainWindow : Window
{
        public MainWindow()
        {
            InitializeComponent();
            this.MouseMove += MainWindow_MouseMove;
        }
        void MainWindow_MouseMove(object sender, MouseEventArgs e)
        {
            PointHitTestResult hit = VisualTreeHelper.HitTest(this, e.GetPosition(this)) as
                    PointHitTestResult;
            if (hit!= null && hit.VisualHit!= null)
            {
                _tooltip.Content = hit.VisualHit.ToString();
                _tooltip.PlacementRectangle = new Rect(e.GetPosition(this), new Size(0, 0));
            }
        }
}
```

运行完整的代码，页面显示效果如图 4.43 所示。其中，CS 代码中的 hit 是指鼠标停留数据。

图 4.42　ToolTip 演示

图 4.43　ToolTipService 显示工具提示信息

4.7.2　Thumb

1. 拖动控件

Thumb 是拖动标，指示一个能被移动的区域。Thumb 作为拖曳对象，一般设置在 Canvas 容器内。实现拖曳只需设置 DragDelta 和 DragStarted 两个事件即可，其中，DragStarted 设置 Thumb 拖动开始状态。

```
<!-- 保留 Window 代码部分 -->
  Title = "窗体拖动" Height = "350" Width = "525">
    < Canvas Background = "LightGreen">
        < Thumb
            Name = "_thum1" Width = "20" Height = "15" Background = "Purple"
            Canvas.Top = "101" Canvas.Left = "116"
            DragStarted = "ThumbStart" DragDelta = "ThumbMove">
        </Thumb>
    </Canvas>
</Window>
```

编写 DragStarted 表示拖动开始事件和 DragDelta 的拖动事件，CS 代码如下。

```
public partial class MainWindow : Window
```

```csharp
{
    public MainWindow()
    {
        InitializeComponent();
    }
    double _startLeft;
    double _startTop;
    private void ThumbStart(object sender,
        System.Windows.Controls.Primitives.DragStartedEventArgs e)
    {
        _startLeft = Canvas.GetLeft(_thum1);
        _startTop = Canvas.GetTop(_thum1);
    }
    private void ThumbMove(object sender,
        System.Windows.Controls.Primitives.DragDeltaEventArgs e)
    {
        double left = _startTop + e.HorizontalChange;
        double top = _startTop + e.VerticalChange;
        Canvas.SetLeft(_thum1,left );
        Canvas.SetTop(_thum1,top);
    }
}
```

运行完整的代码,页面显示效果如图 4.44 所示。DragDelta 为拖动事件,其中拖动事件参数是 System.Windows.Controls.Primitives.DragDeltaEventArgs。获得其拖动从开始发生到当前位移的水平方向变化数值是通过 e.HorizontalChange,获得垂直方向变化数值是通过 e.VerticalChange,它们与显示坐标一致。

图 4.44 Thumb 页面效果

2. 拖动窗口

在窗体的操作中,当把鼠标放在窗口的标题栏上按下就可以拖动窗体。在此,要在窗口的全部位置或特定位置按下鼠标左键实现拖动。在 WinForm 中,要获取鼠标的位置信息,从而设置窗体的位置。当然,WPF 也可以采用 WinForm 类似的方法,但是使用 Thumb 时,操作方法变得简捷了许多。

实现拖动窗口功能,使用 Grid 布局窗口,其 Grid 内部放置一个 Canvas,XAML 代码如下。

```xml
<!-- 保留 Window 代码部分 -->
Title = "窗体拖动" Height = "350" Width = "525">
< Grid Background = "LightGreen"
    MouseLeftButtonDown = "Grid_MouseLeftButtonDown">
< Canvas Name = "Canvas1" Background = "Purple" HorizontalAlignment = "Left"
        Margin = "110,119,0,0" VerticalAlignment = "Top" Height = "55" Width = "83"
        MouseLeftButtonDown = "Canvas1_MouseLeftButtonDown">
    </Canvas >
    </Grid >
</Window>
```

要在 Grid 或 Ganvas 内按下鼠标左键实现窗体拖动，CS 代码如下。

```
public partial class MainWindow : Window
{
    public MainWindow()
    {
        InitializeComponent();
    }
    private void Canvas1_MouseLeftButtonDown(object sender, MouseButtonEventArgs e)
    {
        base.DragMove();
    }
    private void Grid_MouseLeftButtonDown(object sender, MouseButtonEventArgs e)
    {
        base.DragMove();
    }
}
```

运行完整的代码，页面显示效果如图 4.45 所示。DragMove()方法仅用来实现窗体的拖动，在窗体中按下鼠标左键实现"窗体拖动"效果。

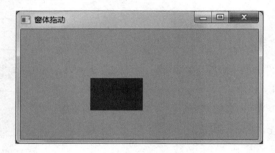

图 4.45 Thumb 拖动窗体的页面效果

4.7.3 Border

Border(边框)是一个装饰性控件，可把 Border 看作能包含子对象的矩形。在边框四周设置不同的厚度和转角半径。设置不同样式的边框，XAML 代码如下。

```
<!-- 保留 Window 代码部分 -->
<Canvas>
    <Border
        Canvas.Left = "12" Canvas.Top = "5"
        BorderThickness = "5"
        CornerRadius = "0"
        BorderBrush = "Green" Padding = "10">
        <TextBox>WPF Border</TextBox>
    </Border>
    <Border
        Canvas.Left = "135" Canvas.Top = "5"
        BorderThickness = "5,3,5,3"
```

```
            CornerRadius = "0,18,0,18"
            BorderBrush = "Green" Padding = "10">
            <TextBox> WPF Border </TextBox>
        </Border>
        <Border
            Canvas.Left = "249" Canvas.Top = "5"
            BorderThickness = "5"
            CornerRadius = "8"
            BorderBrush = "Green" Padding = "10">
            <TextBox> WPF Border </TextBox>
        </Border>
        <Border
            Canvas.Left = "365" Canvas.Top = "5"
            BorderThickness = "15,5,15,5"
            CornerRadius = "16"
            BorderBrush = "Green" Padding = "10">
            <TextBox> WPF Border </TextBox>
        </Border>
    </Canvas>
</Window>
```

运行上述代码，页面显示效果如图 4.46 所示。因为大部分呈现控件（Ellipse、Rectangle 等）不支持子对象，所以运用 Border 控件绘制边框各种样式。尽管在 Border 中只能有一个子控件，但它的子控件中可以包含多个子控件。本例中用到的 Border 属性的含义介绍如下。

图 4.46　不同边缘厚度及转角半径的 Border

BorderBrush 用于设置边框颜色；BorderThickness 用于设置边框的宽度；CornerRadius 用于设置每一个角圆的弧度；Padding 用于设置 Border 中的内容与边框之间的间隔。

4.7.4　Popup

1. 弹出框

WPF 的 Popup（弹出框）用于创建浮动窗口，创建弹出框的 XAML 代码如下。

```
<!-- 保留 Window 代码部分 -->
    <StackPanel>
        <Button Click = "PopupDisplay"> Popup </Button>
        <Popup Name = "_popup" PopupAnimation = "Fade" Placement = "Mouse">
            <Button> Hello, I am a WPF Popup </Button>
        </Popup>
    </StackPanel>
</Window>
```

在 CS 代码中添加事件如下。

```
private void PopupDisplay(object sender, RoutedEventArgs e)
{
    _popup.IsOpen =! _popup.IsOpen;
}
```

运行上述代码,页面显示效果如图 4.47 所示。代码中的"Placement＝"Mouse""是指按下鼠标,响应的行为。

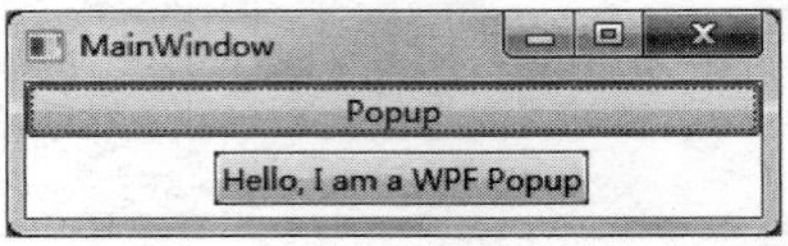

图 4.47 Popup 显示页面

2. 弹出式菜单

在 4.2.1 节中,提到还有一种菜单称为上下文菜单(ContextMenu),又称为弹出式菜单,因为弹出式菜单与弹出框在显示页面上有一定的相似性,故把 ContextMenu 放在了本节来学习。创建弹出式菜单的 XAML 代码如下。

```
<Grid Height = "311" Width = "498">
    <Button x:Name = "cmd" Content = "Popup" Margin = "12,12,401,271">
        <Button.ContextMenu>
            <ContextMenu x:Name = "menu">
                <MenuItem Header = "MenuItem0">
                    <MenuItem Header = "MenuItem00"/>
                    <MenuItem Header = "MenuItem01"/>
                    <MenuItem Header = "MenuItem02"/>
                </MenuItem>
                <MenuItem Header = "MenuItem1"/>
            </ContextMenu>
        </Button.ContextMenu>
    </Button>
</Grid>
</Window>
```

在 CS 代码中添加窗体加载时,响应的事件如下。

```
public partial class MainWindow : Window
{
    public MainWindow()
    {
        InitializeComponent();
    }
    private void Window_Loaded(object sender, RoutedEventArgs e)
    {
        cmd.Click += (obj, args) =>{menu.IsOpen = true; };
    }
}
```

运行上述代码,页面显示效果如图 4.48 所示,在 Button 上右击,弹出一级菜单,将鼠标放在一级菜单上显示二级菜单。

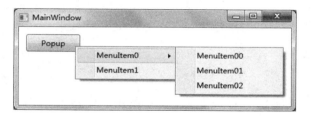

图 4.48 ContextMenu 显示页面

4.7.5 ScrollViewer

ScrollViewer(滚动查看器)可用在需要滚动的任何位置。使用查看器的 XAML 代码如下。

```
<ScrollViewer
    HorizontalScrollBarVisibility = "Auto"
    VerticalScrollBarVisibility = "Auto">
    <Grid>
        <Grid.RowDefinitions>
            <RowDefinition Height = "Auto"/>
            <RowDefinition Height = "Auto"/>
            <RowDefinition Height = "Auto"/>
        </Grid.RowDefinitions>
        <TextBox Margin = "3" Grid.Row = "0" Name = "_toAdd"/>
        <Button Margin = "3" Grid.Row = "1" Click = "AddItem"> Add Name </Button>
        <ListBox Margin = "3" Grid.Row = "2" Name = "_list"/>
    </Grid>
</ScrollViewer>
</Window>
```

在 CS 代码中,为 ListBox 添加数据的方法如下。

```
private void AddItem(object sender, RoutedEventArgs e)
    {
        _list.Items.Add(_toAdd.Text);
    }
```

运行上述代码,页面显示效果如图 4.49 所示。因为列表框本身就带有了滚动查看器,故页面效果与预期不符,ScrollViewer 对布局影响很大,所以需多做案例,在充分理解的基础上,善用它。若想解决上述问题,还可以使用 Viewbox 控件。

4.7.6 Viewbox

Viewbox(视图框)带有一个单独子对象,内容可以随着控件中的文字大小进行缩放。创建 Viewbox 的 XAML 代码如下。

```
<Grid>
    <Viewbox VerticalAlignment = "Top">
        <TextBlock Text = "WPF Viewbox.TextBlock" VerticalAlignment = "Top"/>
```

```
            </Viewbox>
            < Viewbox VerticalAlignment = "Bottom">
                < Button Content = "WPF Viewbox.Button" VerticalAlignment = "Bottom"/>
            </Viewbox>
</Grid>
```

运行上述代码,页面显示效果如图4.50所示。Grid中包含了两个Viewbox,每个Viewbox下包含一个子对象。

图4.49　ScrollViewer显示页面

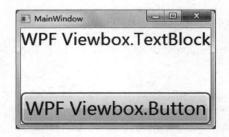

图4.50　Viewbox显示页面

4.8　日　期　控　件

WPF添加了Calendar和DatePicker两个日期控件,这两个控件为用户提供日期选择。
Calendar控件显示日历与Windows操作系统中的日历相似。DatePicker控件与简单的文本框相似,占用的空间少。Calendar和DatePicker公共属性如表4.4所示。

表4.4　Calendar和DatePicker的公共属性

属　性　名	属　性　描　述
DisplayDateStart 和 DisplayDateEnd	日期显示的起始时间和结束时间
BlackoutDates	保存日历中将被禁用或不能选择的日期集合。BlackoutDates.AddDatesInPast()方法:阻止选择被禁用的日期
SelectedDate	为DateTime对象提供选择的日期,没有日期被选中时使用null
DisplayDate	使用DateTime对象确定日历视图中最初显示的日期。若DisplayDate和SelectedDate均为空,使用当前日期
FirstDayOfWeek	默认情况下,日历每周的第一天为周日
IsTodayHighlighted	确定日历视图是否通过突出显示当前日期

4.8.1　Calendar

Calendar(日历)控件除了前面的公共属性外,还有其独有的DisplayMode属性,用来调整日历显示模式,分为Year、Month和Decade 3种。默认情况下为Month模式,显示一个月份,通过单击箭头按钮,选择月份。设置这3种模式,其效果分别如图4.51～图4.53所示。

第 4 章 控件

图 4.51 以月为单位的日历　　　　图 4.52 以日为单位的日历

设置 Calendar 相关属性，自定义日期样式，XAML 代码如下。

```xml
<!-- 保留 Window 代码部分 -->
    Title = "MainWindow" Height = "350" Width = "525">
    <Window.Resources>
        <Style TargetType = "CalendarDayButton" x:Key = "CalendarDayBtnStyle">
            <Setter Property = "Background">
                <Setter.Value>
                    <LinearGradientBrush StartPoint = "0,0" EndPoint = "1,1">
                        <GradientStop Color = "White" Offset = "0"/>
                        <GradientStop Color = "LightBlue" Offset = "1"/>
                    </LinearGradientBrush>
                </Setter.Value>
            </Setter>
        </Style>
    </Window.Resources>
    <Grid>
        <Calendar Margin = "0,50,0,0" Name = "calendarCtl" DisplayMode = "Month"
                  FirstDayOfWeek = "Monday" FlowDirection = "RightToLeft"
                  CalendarDayButtonStyle = "{StaticResource CalendarDayBtnStyle}"/>
    </Grid>
</Window>
```

运行上述代码，页面显示效果如图 4.54 所示。本例中，"FlowDirection＝"RightToLeft""查看日历的方向（从右向左），"FirstDayOfWeek＝"Monday""设置星期一是每行的开始。

图 4.53 以年为单位的日历　　　　图 4.54 日期按键自定义样式

4.8.2 DatePicker

DatePicker（日期选择器）控件，与 Calendar 相比，DatePicker 多出了一个 TextBox 用来

提取从 Calendar 中选择的日期。该文本框以长日期格式或短日期格式保存一个日期字符串。单击 DatePicker 控件的下拉箭头,会弹出与 Calendar 控件相同的日历视图,看起来像下拉组合框。

除了上面的公有属性外,其特有的属性是 IsDropDownOpen,表示是否打开 DatePicker 控件中的下拉日历视图;SelectedDateFormat 是日期格式。

设置 DatePicker 的 SelectedDate 属性为当前日期,XAML 代码如下。

```
<!-- 保留 Window 代码部分 -->
xmlns:sys = "clr-namespace:System;assembly=mscorlib"
Title = "MainWindow" Height = "350" Width = "525">
    <Grid>
        <DatePicker SelectedDate = "{x:Static sys:DateTime.Now}" />
    </Grid>
</Window>
```

运行上述代码,页面显示效果如图 4.55 所示。需要注意的是,XAML 代码中新添加的名称空间。

图 4.55　DatePicker 默认样式

4.9　按　　钮

对于按钮,最熟悉的就是 Button,除此之外,还有 CheckButton 和 RadioButton。

按钮就是响应用户单击的控件,其中按钮最常用的事件就是 Click,它继承自 ButtonBase。Button 实现了标准按钮的默认外观。CheckButton 和 RadioButton 是用于切换开关的按钮,支持 IsChecked(数据模型)和 IsThreeState(交互模型)属性。当 IsThreeState 为 True 时,复选框中的值可从 Checked、Unchecked 和 Indeterminate 这 3 个值中选择。

设置它们的属性,对比三者的外观,XAML 代码如下。

```
<!-- 保留 Window 代码部分 -->
    <Grid>
        <Button Margin = "12,21,0,236" Height = "60"
             HorizontalAlignment = "Left" Width = "142">Button</Button>
        <CheckBox IsChecked = "{x:Null}"
             Margin = "174,21,210,0" Width = "117" Height = "15"
             VerticalAlignment = "Top">Check Null</CheckBox>
        <CheckBox IsChecked = "True"
             Margin = "177,40,210,256">Check True</CheckBox>
        <CheckBox IsChecked = "False"
             Margin = "177,60,210,236">Check False</CheckBox>
        <RadioButton IsChecked = "{x:Null}"
             Margin = "321,21,12,275">RadioButton Null</RadioButton>
        <RadioButton IsChecked = "True"
             Margin = "321,40,12,256">RadioButton True</RadioButton>
        <RadioButton IsChecked = "False"
             Margin = "321,60,12,236">RadioButton False</RadioButton>
    </Grid>
</Window>
```

运行上述代码,页面显示效果如图 4.56 所示。这些控件看上去就像内置的 Windows 控件,页面还显示了其所有交互逻辑。对于这些基础的控件,WPF 就是一个内置控件外观的完善复制。

图 4.56　几种按钮的默认样式

4.10　小　　结

本章介绍了 WPF 控件的新概念——内容模型和模板,并了解元素合成、富内容和简单的编程模型的控件原则。在此基础上,学习了 WPF 的内置控件。

读者学完前 4 章,已经掌握了有关 WPF 基础的编程内容和界面 UI,现在可以尝试并做出赏心悦目的界面。从第 5 章将开始 WPF 的高级进阶。

习题与实验 4

1. 对比 ListBox 和 ComboBox 的异同点及使用场景。
2. 设计搜索界面,在页面底部显示日期及时间,如图 4.57 所示。

图 4.57　搜索页面

3. 用 Grid 布局登录页面,拖动窗体右下角调整窗体大小时,控件会随着窗体的变化,按比例调整自身的大小。页面显示效果如图 4.58 所示。

图 4.58　Grid 布局登录页面

为了便于读者看清楚该页面的 Grid 布局方式，给出页面的 Logical Tree，如图 4.59 所示。

图 4.59　Grid 布局登录页面的逻辑结构

4. 设计连连看游戏的页面如图 4.60 所示。

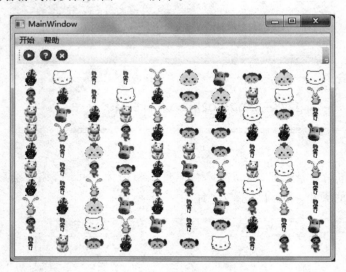

图 4.60　连连看布局

第 5 章

数据

从早期的控制台用户界面(Control User Interface,CUI)到图形用户界面(Graphic User Interface,GUI)的发展,计算机操作系统经历了 DOS(Disk Operating System)到 Windows 的飞跃,带给用户良好的视觉体验。在 Windows 操作系统的图形用户界面中,用户通过单击图标,控件启动应用程序,把这动作可以抽象成为 UI(控件)驱动程序的思想。在 WPF 中,实现了数据驱动 UI 的设计思想,数据成为应用程序的中心。

5.1 数据驱动模型

数据模型是数据源与数据调用者之间的约定。在微软发展历程中,每个框架都具有自己的数据模型。例如,Visual Basic 具有 DAO 数据模型(Data Access Objects,数据访问对象模型)、RDO(Remote Data Objects,远程数据对象模型)和 ADO(ActiveX Data Objects,ActiveX 数据对象)。在.NET 中,把 API 的数据模型作为整个框架的数据模型。

5.1.1 数据原则

绝大多数应用程序都涉及数据的创建、编辑和显示。无论数据是以数字、文本、图形或 3D 等形式出现,应用程序都要对其进行操作。当然,有诸多方式来表示数据。在.NET 中使用标准的数据模型增强了框架处理数据的能力。

1..NET 数据模型

.NET 数据模型支持基础的数据模型,还支持类、接口、结构、枚举和委托类型。在.NET 中,System.Collection 的接口可以表示列表,如 IEnumerable、IList。属性可通过内置的 CLR 属性来定义,也可以通过 ICustomTypeDescriptor 实现。现在新的数据访问技术 LINQ(Language Intergrated Query,语言集成查询)可以在.NET 中使用,此时基础数据模型保持不变。

.NET 数据处理方式有 ADO.NET(位于 System.Data 名称空间)、XML(位于 System.Xml 名称空间)、数据契约(位于 System.Runtime.Serialization 名称空间)和标记(位于 System.Windows.Markup 名称空间)。众多的.NET 数据处理方式都是构建在基础的数据处理模型之上的。

因为所有的 WPF 数据操作都是基于基本的.NET 数据模型,所以 WPF 控件可以接收任意 CLR 对象数据。

```
ListBox lb = new ListBox();
lb.ItemsSource = new string[] {"Hello","Good Morning","Bye"};
```

只要数据通过 CLR 方式访问,在 WPF 中就可以被显示。

2. 绑定的编程方式

WPF 最显著的特点就是绑定的编程方式集成到整个系统。例如,控件模板是模板绑定;资源通过绑定来加载;Button 控件是基于内容模型,使用数据绑定。

不同的绑定名称对应不同的绑定功能与场景。例如,标准绑定是{Binding}标记;模板绑定是{TemplateBinding}标记,只用于模板的上下文中,只能绑定到模板控件的属性上。大多数控件可以绑定到模板上。

绑定的目的是当数据发生变化时,让两个对象保持同步。若两个对象的数据类型不能完全匹配,则需数据转换。

3. 数据转换

既然在 WPF 中,绑定的编程方式集成到整个系统。这意味着数据无论绑定到哪儿,都可以被完全转换。WPF 对数据转换的规定是:除了绑定在资源的数据不支持转换,其他都可以被完全转换。因为资源是用于呈现的,无须转换。

WPF 包含值转换和数据模板两种类型转换。其中,值转换又分为单值转换接口和多值转换接口。数据模板则允许控件被动态地创建出来,表示数据。

因为资源是程序中最常见的数据源,所以下面将以资源开始学习 WPF 中的数据。

5.1.2 资源

在第 1 章的学习中就已经出现了资源。在计算机程序中,只要是对程序有用的对象,都可以统称为资源。

在窗体中的 Resources 属性中,定义一个笔刷,设置 Button 中的背景色与笔刷中定义的颜色相同,XAML 代码如下。

```
<!-- 保留 Window 代码部分 -->
    Title = "MainWindow" Height = "350" Width = "525">
    < Window.Resources >
        < SolidColorBrush x:Key = "toColor">LightGreen</SolidColorBrush >
    </Window.Resources >
    < Button Background = "{StaticResource toColor}"
            Foreground = "Red" Margin = "30" >LightGreen Button</Button >
</Window >
```

运行上述代码后,页面显示效果如图 5.1 所示。由于笔刷为 LightGreen,Button 中的 Background 属性以静态资源的方式通过键名调用笔刷,于是 Button 的背景色也是 LightGreen(亮绿色)。

接下来在资源字典(ResourceDictionary)中定义 TextBlock 的 Text 属性绑定静态资源,XAML 代码如下。

```
<!-- 保留 Window 代码部分 -->
        xmlns:sys = "clr - namespace:System;assembly = mscorlib"
        Title = "MainWindow"   Height = "350" Width = "525">
    < Window.Resources >
        < ResourceDictionary >
            < sys:String x:Key = "str">夜来风雨声,花落知多少.</sys:String >
```

```
            </ResourceDictionary>
        </Window.Resources>
        <StackPanel>
            <TextBlock Text = "{StaticResource ResourceKey = str}"
                       Margin = "30" Width = "220" FontSize = "18"/>
        </StackPanel>
    </Window>
```

运行上述代码后,页面显示效果如图 5.2 所示。注意在 XAML 中声明了 CLR 新的名称空间后,再在数据字典中定义了字符串,并命名为 str,在 TextBlock 的 Text 属性,绑定静态资源后,通过"ResourceKey=str"设置后,方能正确地访问资源字典。

图 5.1 窗体中定义笔刷资源

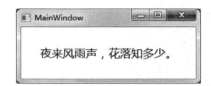
图 5.2 资源字典

在 WPF 中,界面上的每一个元素都是一个对象,并且都有一个名为 Resources 的属性, Resources 属性继承于 FrameworkElement 类,其类型为 ResourceDictionary。由于 Resources 是复数形式,因此每一个对象可以拥有多个资源。由于资源的多样化,故获取到的资源的类型为 Object 类型。在获取到资源时,必要时要转化成符合要求的类型。资源与数据结构中的哈希表相仿,对象都是以键值对(Key-Value)的形式相关联,当需要某个资源时,可以通过 Key-Value 来索引。

资源是数据绑定的一种特殊形式,它优化了数据绑定(这里的数据绑定更新较少)。在 WPF 中,常见的数据绑定名称是 Binding。本书还将在第 10 章讲解 WPF 中的对象级资源和二进制资源。

5.2 数据绑定原理

数据绑定是在应用程序前台 UI 表现层与后台业务逻辑层之间搭建的一座桥梁,实现数据同步更新。那么数据绑定这座桥梁则连接着数据源和数据目的地。通常,将从数据源到数据目标通道称为路径(Path)。

5.2.1 数据绑定机制

首先来看一个示例,页面上包含用户信息中的姓名与年龄,放一个命令按钮实现年龄加 1 的操作。其对应的 XAML 代码如下。

```
<!-- 保留 Window 代码部分 -->
        Title = "MainWindow" Height = "175" Width = "230">
    <Grid Name = "grid" Height = "307" Width = "450">
        <TextBlock Height = "23" HorizontalAlignment = "Left"
                   Margin = "12,12,0,0" Text = "Name:" VerticalAlignment = "Top" />
```

```
            <TextBlock Height = "23" HorizontalAlignment = "Left"
                 Margin = "12,0,0,237" Text = "Age:" VerticalAlignment = "Bottom" />
            <TextBox Height = "23" HorizontalAlignment = "Left" Width = "120"
                 Margin = "74,12,0,0" Name = "nameTextBox" VerticalAlignment = "Top" />
            <TextBox Height = "23" HorizontalAlignment = "Left" Width = "120"
                 Margin = "74,51,0,0" Name = "ageTextBox" VerticalAlignment = "Top" />
            <Button Content = "Age++" Height = "23" HorizontalAlignment = "Left" Width = "182"
                 Margin = "12,98,0,0" Name = "ageaddButton" VerticalAlignment = "Top" />
    </Grid>
</Window>
```

运行上述代码,页面显示效果如图 5.3 所示。在 XAML 页面上放置两个 TextBlock(用于用户提示信息)、两个 TextBox(接收姓名与年龄)、一个 Button(实现年龄加 1 操作)。

接下来在后台 CS 代码中创建一个用户类(User),用户类含有 Name 和 Age 两个属性。代码如下。

```
public class User{
    string name;
    public string Name
    {
        get { return this.name; }
        set { this.name = value; }
    }
    int age;
    public int Age
    {
        get { return this.age; }
        set { this.age = value; }
    }
    public User() { }
    public User(string name, int age)
    {
        this.name = name;
        this.age = age;
    }
}
```

在此,定义与前台相对应的 User 类及其属性,使用类的目的:在代码中不直接操作控件,增强程序的可维护性。

在后台 CS 代码中加入对象实例化,为对象赋初值,设置 grid 的数据上下文。

```
public partial class MainWindow : Window
{   User user = new User("谢浩然", 9);
        public MainWindow()
        {   InitializeComponent();
            //Let the grid know its data context
            grid.DataContext = user;
            this.nameTextBox.Text = user.Name;
            this.ageTextBox.Text = user.Age.ToString() ;
        }
}
```

下面需要在 XAML 页面上,将 User 类中的 Name 和 Age 两个属性分别绑定到 nameTextBox 与 ageTextBox 两个控件的 Text 属性,在 XAML 页面中需增添的代码如下。

```
<!-- 保留 Window 代码部分 -->
xmlns:local = "clr-namespace:DataBindingMode"
    Title = "MainWindow" Height = "175" Width = "230" >
  < Grid Name = "grid" Height = "307" Width = "450" >
    ...
  < TextBox Text = "{Binding Path = Name}" Name = "nameTextBox" />
  < TextBox Text = "{Binding Path = Age}" Name = "ageTextBox" />
```

运行完整的代码,页面显示效果如图 5.4 所示。页面中的 Name 与 Age 与后台 CS 代码中设置的值相同。此页面上,数据绑定正确无误。

图 5.3　未实现数据绑定的 XAML 页面　　　图 5.4　XAML 页面属性绑定后台数据

XAML 页面中的 Age++ 按钮功能是:单击该按钮,年龄加 1,并在弹出对话框中显示当前年龄加 1 后的年龄值。先在前台 XAML 页面中添加 ageaddButton 按钮的 ageaddButton_Click 事件,在后台用 CS 编写按钮功能,代码如下。

```
private void ageaddButton_Click(object sender, RoutedEventArgs e)
{   ++user.Age;       //user_PropertyChanged will update ageTextBox
    MessageBox.Show(string.Format("Happy Birthday,{0}, Age: {1}",
    user.Name, user.Age), "Birthday");
}
```

运行上述代码,单击 Age++ 按钮后,页面显示效果如图 5.5 所示。弹出框中的年龄是 10,但是 XAML 页面中的 Age 还是 9。可见,后台数据已正确更新,但是更新后的数据未传回前台页面。

重新运行代码,在 XAML 页面上输入数据后,单击 Age++ 按钮,页面显示效果如图 5.6 所示。从页面上可知,前台 XAML 页面中的数据更新后,后台 CS 代码中数据实现了实时更新。

图 5.5　实现 Age++ 按钮的功能的页面数据　　　图 5.6　XAML 页面更新对应后台数据

对比图 5.5 与图 5.6 的页面可知,在前台 XAML 页面中的数据更新,后台同步实时更新了数据。但是当后台 CS 代码中的数据更新时,前台 XAML 页面的数据未能更新。这种数据绑定机制可理解为单向绑定机制。XAML 页面数据变化能传到后台 CS 中的原因是:每个 TextBox 都有自己的 TextChanged 事件实现同步更新。由此可以推理,当后台数据变更时,则应设置监听事件,实现实时更新前台数据。接下来重新编写类的定义,CS 代码如下。

```
using System.ComponentModel;        //INotifyPropertyChanged 的 namespace
…
public class User:INotifyPropertyChanged{
    public event PropertyChangedEventHandler PropertyChanged;
    protected void Notify(string propName){
        if (this.PropertyChanged!= null) {
            PropertyChanged(this, new PropertyChangedEventArgs(propName));
        }
    }
    string name;
    public string Name
    {
        get { return this.name; }
        set {
            if (this.name == value) { return; }
            this.name = value;
            Notify("Name");
        }
    }
    int age;
    public int Age
    {
        get { return this.age; }
        set
        {
            if (this.age == value) { return; }
            this.age = value;
            Notify("Age");
        }
    }
    public User() { }
    public User(string name, int age)
    {
        this.name = name;
        this.age = age;
    }
}
```

编写 user_PropertyChanged 事件的 CS 代码如下。

```
void user_PropertyChanged(object sender, PropertyChangedEventArgs e)
    {
        switch (e.PropertyName)
```

```
            {
                    case "Name": this.nameTextBox.Text = user.Name;break;
                    case "Age": this.ageTextBox.Text = user.Age.ToString(); break;
            }
    }
```

在 Window 中需要监听属性变化，需添加监听语句如下。

```
public MainWindow()
{
            InitializeComponent();
            grid.DataContext = user;
            this.nameTextBox.Text = user.Name;
            this.ageTextBox.Text = user.Age.ToString();
        //Watch for changes in user's properties
            user.PropertyChanged += user_PropertyChanged;
}
private void ageaddButton_Click(object sender, RoutedEventArgs e)
{
            ++user.Age;            //user_PropertyChanged will update ageTextBox
            MessageBox.Show (string.Format("Happy Birthday,{0},Age:{1}",user.Name,
                    user.Age),"Birthday");
}
```

运行上述代码，单击 Age++ 按钮后，看到前台 XAML 页面与弹出框内容显示页面如图 5.7 所示。在前台 XAML 页面输入数据，后台数据与前台数据同步更新，如图 5.8 所示。

图 5.7　后台数据更新与前台页面同步

图 5.8　前台页面更新与后台数据同步更新

INotifyPropertyChanged 是系统的接口，实现后台对象的值发生改变前台页面实时更新的功能。注意添加系统引用，实现接口的调用。在 5.3 节中重点介绍该接口的用法。

在项目开发时，根据需求，实现单向或双向的绑定机制。WPF 中的数据绑定机制用于解决数据同步更新问题。

5.2.2　数据源与路径

绑定的目的是保持相关联的数据点之间的数据同步。这个数据点既包含了数据的出发点，也包含了数据的目的地。数据点可以看作单独的"结点"。要描述数据点，通常要借助数据源和路径（查询）。

在 XAML 页面中添加 TextBox 与 ContentControl 两个控件，它们的名字分别是 textBox1、contentControl1。

在 WPF 中,Binding 类就表示了数据点,创建一个引用 TextBox 对象 Text 属性的数据点。在后台用 CS 代码来构造绑定,需要有数据源和路径(查询),代码如下。

```
Binding bind = new Binding();
bind.Source = textBox1;
bind.Path = new PropertyPath("Text");
```

接下来构建 contentControl1 数据点,让它和 textBox1 数据点保持同步。WPF 的数据绑定是指将数据绑定到元素树或者数据来自元素树。在此,先用 SetBinding 方法定义 contentControl1 数据点。

```
contentControl1.SetBinding(ContentControl.ContentProperty,bind);
```

上面是在后台用 CS 代码构建数据点,并设置路径(查询),进一步理解数据源和路径。接下来用 XAML 代码来实现数据绑定,代码如下。

```
<!-- 保留 Window 代码部分 -->
<Grid>
    <TextBox Height = "23" HorizontalAlignment = "Left" Margin = "10,10,0,0"
             Name = "textBox1" VerticalAlignment = "Top" Width = "168" />
    <ContentControl Height = "50" Margin = "10,58,0,0" Name = "contentControl1"
                    Content = "{Binding ElementName = textBox1,Path = Text}"
                    VerticalAlignment = "Top" />
</Grid>
</Window>
```

运行上述代码,在 TextBox 中输入字符串时,ContentControl 控件中的值与之相同。页面显示效果如图 5.9 所示。

将 TextBox 中的 Text 属性(字符串)绑定到 Content 属性(对象),还可以将 Text 属性绑定到 FontFamily 属性(它并不是一个字符串),其 XAML 代码如下。

```
<!-- 保留 Window 代码部分 -->
<StackPanel>
    <TextBox x:Name = "textBox1"/>
    <TextBox x:Name = "textBox2"/>
    <ContentControl Margin = "8" Content = "{Binding ElementName = textBox1,Path = Text}"
                    FontFamily = "{Binding ElementName = textBox2,Path = Text}"/>
</StackPanel>
</Window>
```

运行上述代码,页面显示效果如图 5.10 所示,表示一个字符串转换为 FontFamily。

图 5.9 TextBox 绑定 ContentControl

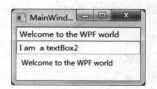
图 5.10 TextBox 绑定到 FontFamily

5.2.3 值转换机制

值转换的基础机制有 TypeConverter 和 IValueConverter。其中,TypeConverter 是在 .NET1.0 版本中就具有的基础机制。在上节中出现的 FontFamily 是使 TypeConverter 与 FontFamily 发生关联,系统自动发生转换。IValueConverter 是 WPF 独有的值转换器。

在 WPF 中,通过绑定关联的值转换器可以把一种类型转换成任何类型(内置类型和自定义类型)。接下来在后台创建用户自定义类型,CS 代码如下。

```
public class Person
{
    private string name;
    public string Name {
        get{return name;}
        set{name = value;} }
}
```

在此,把 Text 类型转换成特殊的类型,需要编写转换器,继承自 IValueConverter。因为 IValueConverter 接口还包含 Convert 和 ConvertBack 方法的定义,所以值转换器都必须实现 Convert 和 ConvertBack 两种方法。代码如下。

```
public class PersonConverter : IValueConverter {
    public object Convert(object value, Type targetType, object parameter, System.Globalization.CultureInfo culture)
    {
        Person p = new Person();
        p.Name = (string)value;
        return p;
    }
    public object ConvertBack(object value, Type targetType, object parameter,
                    System.Globalization.CultureInfo culture)
    {
        return ((Person)value).Name;
    }
}
```

上述代码中的 Convert 方法表示从绑定数据起点到绑定目标的值转换,ConvertBack 方法表示从绑定目标到绑定数据起点的值转换。因此,若绑定模式是一次性绑定或单向绑定,只要在 Convert 方法的实现中完成值转换的工作即可;如果是双向绑定,就要同时在 Convert 和 ConvertBack 两种方法中完成值转换的工作。

使用值转换的最后一步是与控件属性绑定,XAML 绑定代码如下。

```
<Window x:Class = "ConverterDemo.MainWindow"
        xmlns = "http://schemas.microsoft.com/winfx/2006/xaml/presentation"
        xmlns:x = "http://schemas.microsoft.com/winfx/2006/xaml"
        xmlns:local = "clr-namespace:ConverterDemo"
        Title = "MainWindow" Height = "350" Width = "525">
    <StackPanel>
        <TextBox x:Name = "textBox1"/>
        <TextBox x:Name = "textBox2"/>
        <ContentControl Margin = "8" Height = "117"
```

```xml
                    FontFamily = "{Binding ElementName = textBox2, Path = Text}">
            <ContentControl.Content>
                <Binding ElementName = "textBox1" Path = "Text">
                    <Binding.Converter>
                        <local:PersonConverter
                            xmlns:local = "clr-namespace:ConverterDemo">
                        </local:PersonConverter>
                    </Binding.Converter>
                </Binding>
            </ContentControl.Content>
        </ContentControl>
    </StackPanel>
</Window>
```

运行上述代码，ContentControl.Content 已是值转换器规定的内容，在 TextBox 中输入字符串后，页面显示效果如图 5.11 所示。需要注意的是，在 XAML 页面调用用户自定义类型时，需要声明，格式如下。

```xml
xmlns:local = "clr-namespace:ConverterDemo"
```

其中，local 为用户自定义字符。

数据模板通过 DataType 属性取得数据块，并构建一棵显示树。接下来为 Person 类构建简单模板，并把数据绑定到 Name 属性上，在上例中的代码的基础上，添加相应的 XAML 代码如下。

```xml
    <DataTemplate xmlns:local = "clr-namespace:ConverterDemo"
                  DataType = "{x:Type local:Person}">
        <Border Margin = "8" Padding = "5" BorderBrush = "Green"
                BorderThickness = "5" CornerRadius = "3">
            <TextBlock Text = "{Binding Path = Name}"/>
        </Border>
    </DataTemplate>
```

把这个模板关联到 ContentControl，设置 ContentTemplate 属性，相应的 XAML 代码如下。

```xml
    <ContentControl Margin = "8" Height = "117"
                    FontFamily = "{Binding ElementName = textBox2, Path = Text}">
        <ContentControl.Content>
            <Binding ElementName = "textBox1" Path = "Text">
                <Binding.Converter>
                    <local:PersonConverter
                        xmlns:local = "clr-namespace:ConverterDemo">
                    </local:PersonConverter>
                </Binding.Converter>
            </Binding>
        </ContentControl.Content>
        <ContentControl.ContentTemplate>
            <DataTemplate xmlns:local = "clr-namespace:ConverterDemo"
                          DataType = " {x:Type local:Person}">
```

```
                    < Border Margin = "8" Padding = "5" BorderBrush = "Green"
                        BorderThickness = "5" CornerRadius = "3">
                        < TextBlock Text = "{Binding Path = Name}"/>
                    </Border >
                </DataTemplate >
            </ContentControl.ContentTemplate >
        </ContentControl >
    </StackPanel >
</Window >
```

运行上述代码,页面显示效果如图 5.12 所示。其中< TextBlock Text = "{Binding Path= Name}"/>设置数据绑定。

图 5.11 值转换器实现绑定

图 5.12 数据模板实现绑定

WPF 数据绑定允许任意数据上下文和元素建立关联。通过使用数据模板和数据上下文可以自动设置为要进行转换的数据。可以在任何元素上显式地设置 DataContext 属性及在元素或它们的子元素上用于绑定的数据源。

在了解数据绑定的基本原理后,需要再学习数据绑定的用法,才能深刻地理解数据模板系统。

5.2.4 数据绑定模型

数据绑定的基本思想是数据点和数据转换。掌握了数据绑定的原理后,首先来认识数据绑定的模型,通过前两节的学习,认识到数据点(包含数据源和数据目标)、路径(path)及数据转换等数据绑定基础知识,此时可以建立数据绑定模型,如图 5.13 所示。

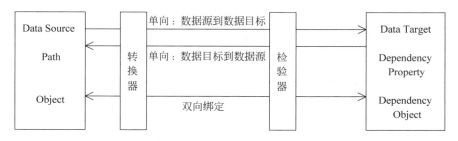

图 5.13 数据绑定模型

由数据绑定模型可知,一个绑定有数据源、路径、数据目标和目标的依赖属性 4 个组件。

尽管从图 5.13 可知,数据绑定模式分成单向和双向两大类。但在 WPF 的 Binding .Mode 属性共有 5 个枚举值,如表 5.1 所示。

表 5.1　Binding.Mode 属性值功能

名　称	功　能
OneWay	当源属性变化时，更新目标属性
TwoWay	当源属性变化时，更新目标属性；当目标属性变化时，更新源属性
OneTime	根据源属性设置目标属性，但不传播后续更改，降低系统开销
OneWayToSource	当目标属性变化时，更新源属性
Default	未指明绑定模式，则依赖于目标属性，可以是双向的（用户可以设置的属性，TextBox.Text 属性），也可以是单向的（对于所有其他属性）

表 5.1 给出了数据绑定模式的 5 种方式。当数据绑定模式未给出时，系统默认 Default 类型。接下来，继续使用数据绑定机制中的案例（XAML 包含姓名、年龄、年龄加 1 按钮），在此基础上，自定义"检验器"。

检验器功能：对用户在 XAML 页面输入的年龄值进行校验，规定年龄值：0～120 的整数。若不在此范围，给出用户出错提示信息。

在后台创建数值校验规则类，命名为 NumberRangeRule。CS 代码如下。

```
public class NumberRangeRule : ValidationRule
{
    int min;
    public int Min
    {
        get { return min; }
        set { min = value; }
    }
    int max;
    public int Max
    {
        get { return max; }
        set { max = value; }
    }
    public override ValidationResult Validate (object value, System.Globalization
                                                .CultureInfo cultureInfo)
    {
        int number;
        if (!int.TryParse((string)value, out number))
        {
            return new ValidationResult(false, "Invalid number format");
        }
        if (number < min || number > max)
        {
            return new ValidationResult (false, string.Format("Number out of range ({0} -
                                        {1})", min, max));
        }
        return ValidationResult.ValidResult;
    }
}
```

由于 NumberRangeRule 继承自系统的 ValidationRule，重载了系统的 Validate() 方法。

在 XAML 页面中需要添加的代码如下。

```xml
<Window x:Class = "DataBindingMode.MainWindow"
        xmlns = "http://schemas.microsoft.com/winfx/2006/xaml/presentation"
        xmlns:x = "http://schemas.microsoft.com/winfx/2006/xaml"
        xmlns:local = "clr-namespace:DataBindingMode"
        Title = "MainWindow" Height = "175" Width = "230" Loaded = "Window_Loaded">
    <!-- 保留原有控件代码部分 -->
    <TextBox Height = "23" Width = "120" HorizontalAlignment = "Left" Margin = "74,51,0,0" Name = "ageTextBox" VerticalAlignment = "Top">
        <TextBox.Text>
            <Binding Path = "Age" NotifyOnValidationError = "True">
                <Binding.ValidationRules>
                    <local:NumberRangeRule Min = "0" Max = "120"
                        xmlns:local = "clr-namespace:DataBindingMode" >
                    </local:NumberRangeRule>
                </Binding.ValidationRules>
            </Binding>
        </TextBox.Text>
        <TextBox.ToolTip>
            <ToolTip>
                <TextBlock Margin = "5">年龄:0~120 的整数</TextBlock>
            </ToolTip>
        </TextBox.ToolTip>
    </TextBox>
    <!-- 保留原有控件代码部份 -->
```

TextBox 的 ToolTip 功能是当鼠标放在文本框位置时,实现界面友好的语义提示。再在 ageaddButton_Click() 中添加一条校验语句如下,该语句位于 ++user.Age;语句之前:

```
Validation.AddErrorHandler(this.ageTextBox, ageTextBox_ValidationError);
```

还需要在后台给 ageTextBox 写一段校验方法,CS 代码如下:

```
void ageTextBox_ValidationError(object sender, ValidationErrorEventArgs e)
{
    MessageBox.Show((string)e.Error.ErrorContent, "Validation Error");
    ageTextBox.ToolTip = (string)e.Error.ErrorContent;
}
```

运行完整的代码,在 Age 对应的文本框中输入"888"后,文本框变红,引起用户注意,如图 5.14 所示。在单击 Age++ 按钮后,弹出有效性校验窗口,如图 5.15 所示。

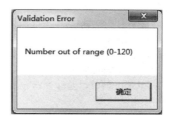

图 5.14 输入不合法的 XAML 页面　　　　图 5.15 有效性校验窗口

在前面的案例中,并未明确给出 Binding.Mode,此时"Binding.Mode=Default"。当目

标属性是 TextBox.Text 时，实现双向绑定。

5.3 数据绑定用法

数据绑定是 WPF 程序重要的新技术，由于其强大的功能及灵活性也导致软件开发人员在编写绑定的过程中产生很大的疑问。本节从数据绑定的习惯用法出发，诠释 WPF 数据驱动 UI 的思想。

5.3.1 控件间的绑定

1. 不同控件间绑定

让 Slider 控件与 TextBox 控件绑定，XAML 代码如下。

```
<Grid>
    <Slider Name="slider1" Height="23" HorizontalAlignment="Left"
            Margin="10,10,0,0" VerticalAlignment="Top" Width="188" />
    <TextBox Name="textBox1" Height="23" HorizontalAlignment="Left"
             VerticalAlignment="Top" Width="188" Margin="10,43,0,0"
             Text="{Binding ElementName=slider1,Path=Value,Mode=TwoWay}">
    </TextBox>
</Grid>
```

运行上述代码，当拖动滑块时文本框中的数值会随着滑块动态改变，如图 5.16 所示。在文本框中输入"3"后，按下 Tab 键（释放焦点），滑块会移到相应位置，如图 5.17 所示。这是"Mode=TwoWay"时，XAML 页面的变化。请读者将绑定模式更改后，查看页面变化。

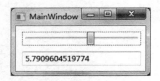
图 5.16 文本框数值随着滑块变化

图 5.17 滑块随着文本框变化

2. 相同控件间绑定

让两个 TextBox 绑定，在第 1 个 TextBox 中输入数值时，第 2 个 TextBox 中同步显示；当在第 2 个 TextBox 中输入数值时，第 1 个 TextBox 也实时更新。XAML 代码如下。

```
<StackPanel>
    <Label>TwoWayInput String</Label>
    <TextBox Name="textBoxTwoWayInput"/>
    <Label>TwoWayOutput String:</Label>
    <TextBox Name="textBoxTwoWayOutput" Text="{Binding Text, ElementName=textBoxTwoWayInput,
             UpdateSourceTrigger=PropertyChanged }" />
</StackPanel>
```

运行上述代码，在名为 textBoxTwoWayInput 的 TextBox 中输入"I am Input"，如图 5.18 所示；在名为 textBoxTwoWayOutput 的 TextBox 中输入"I am Output"，如图 5.19 所示。本例与上例的区别是：此处绑定时，UpdateSourceTrigger 属性值设为 PropertyChanged，这时，

TextBox 控件的默认值是 LostFocus(失去焦点)，就可以做到实时更新。

图 5.18　TextBox 间绑定

图 5.19　实时更新 TextBox 绑定值

5.3.2　控件绑定资源文件值

在此，用一个 TextBox 绑定资源文件的值，再加入一个 TextBox，让它与前一个 TextBox 绑定。前台设计页面的 XAML 代码如下。

```
<Window x:Class = "DataBindingUsage0.MainWindow"
     xmlns = "http://schemas.microsoft.com/winfx/2006/xaml/presentation"
     xmlns:x = "http://schemas.microsoft.com/winfx/2006/xaml"
     xmlns:sys = "clr-namespace:System;assembly = mscorlib"
     Title = "MainWindow" Height = "350" Width = "525">
  <Window.Resources>
     <sys:String x:Key = "TestInfo">Hello World!</sys:String>
  </Window.Resources>
  <StackPanel>
     <Label>Read from resources:</Label>
     <TextBox Name = "textBox" Text = "{StaticResource TestInfo}"></TextBox>
     <TextBox Text = "{ Binding Text,ElementName = textBox,
         UpdateSourceTrigger = PropertyChanged}"/>
  </StackPanel>
</Window>
```

运行上述代码，页面显示效果如图 5.20 所示，两个 TextBox 都绑定了资源文件字符。在 TextBox 中输入新的内容后，两个 TextBox 值相同，如图 5.21 所示。

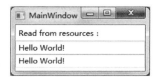
图 5.20　TextBox 绑定资源文件的值

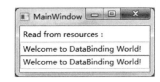
图 5.21　两个 TextBox 联动

5.3.3　属性变更通知接口

数据绑定机制中的案例是为了阐明数据绑定的原理，看起来有些复杂，在实际使用数据绑定时，只要调用 INotifyPropertyChanged(属性变更通知接口)，程序会变得更简洁。本节将重点介绍 INotifyPropertyChanged 的使用方法。

属性变更通知接口的实现和使用步骤如下。

(1) 创建一个类，继承并实现 INotifyPropertyChanged 接口。

(2) 在该类中创建一个名为 Notify() 的包装函数，其功能是实现属性变更通知(也就是说，执行 PropertyChanged 事件，该事件含有一个参数，参数为 string 类型）。

（3）实现属性变更通知。如果属性的实际值发生更改后，则调用 Notify()。

针对数据绑定机制中的案例，先来编写 CS 中的 User 类，代码如下。

```csharp
using System.ComponentModel; //引用 INotifyPropertyChanged
namespace WPF_IPropertyChanged
{//Step 1: User 类继承 INotifyPropertyChanged
    public class User : INotifyPropertyChanged
    {//Step 2: Notify 包装函数定义
        protected void Notify(string propName)
        {
            if (this.PropertyChanged!= null)
            {
                PropertyChanged(this, new PropertyChangedEventArgs(propName));
            }
        }
        string name;
        public string Name
        {
            get { return this.name; }
            set { if (this.name == value) { return;};
            this.name = value;
            Notify("Name");        //Step 3: 实现属性变更通知
            }
        }
        int age;
        public int Age
        {
            get { return this.age; }
            set {if (this.age == value) { return; }
            this.age = value;
            Notify("Age"); //Step 3: 实现属性变更通知
            }
        }
        public User() { }
        public User(string name, int age)
        {
            this.name = name;
            this.age = age;
        }
        public event PropertyChangedEventHandler PropertyChanged;          //Step 1: 实现接口
    }
}
```

在前台页面的 XAML 代码如下。

```xaml
<Window x:Class = "WPF_IPropertyChanged.MainWindow"
    xmlns = "http://schemas.microsoft.com/winfx/2006/xaml/presentation"
    xmlns:x = "http://schemas.microsoft.com/winfx/2006/xaml"
    xmlns:local = "clr-namespace:WPF_IPropertyChanged"
    Title = "MainWindow" Height = "350" Width = "525">
  <Grid Height = "307" Width = "450" Name = "grid">
    <TextBlock Height = "23" HorizontalAlignment = "Left"
        Margin = "12,12,0,0" Text = "Name:" VerticalAlignment = "Top" />
    <TextBlock Height = "23" HorizontalAlignment = "Left"
        Margin = "12,0,0,237" Text = "Age:" VerticalAlignment = "Bottom" />
```

```
            <TextBox Text="{Binding Name}" Height="23" HorizontalAlignment="Left" Width="120"
                Margin="74,12,0,0" Name="nameTextBox" VerticalAlignment="Top" />
            <TextBox Text="{Binding Age}" Height="23" Width="120"
                    HorizontalAlignment="Left" Margin="74,51,0,0" Name="ageTextBox"
                    VerticalAlignment="Top"/>
            <Button Content="Age++" Height="23" HorizontalAlignment="Left" Width="182"
                    Margin="12,98,0,0" Name="ageaddButton" VerticalAlignment="Top" Click
                    ="ageaddButton_Click" />
    </Grid>
</Window>
```

观察上述代码,两个 TextBox 分别只绑定了后台的 Name 和 Age 属性。

下面给出 MainWindow.xaml.cs 的代码。

```
namespace WPF_IPropertyChanged{
    /// <summary>
    /// MainWindow.xaml 的交互逻辑
    /// </summary>
    public partial class MainWindow : Window
    {                              //创建 User 类的对象,并赋初值;
        User user = new User("谢浩然", 9);
        public MainWindow()
        {
            InitializeComponent();
            grid.DataContext = user; //把对象 user 赋值给前台名为 grid 的数据上下文;
            this.nameTextBox.Text = user.Name; //将 user 对象的 Name 属性绑定到前台控件;
            this.ageTextBox.Text = user.Age.ToString(); }
        private void ageaddButton_Click(object sender, RoutedEventArgs e)
        {
            ++user.Age;
            MessageBox.Show (string.Format("Happy Birthday,{0}, Age: êo{1}", user.Name,
                    user.Age), "Birthday");
        }
    }
}
```

运行上述代码,页面显示效果与图 5.7 和图 5.8 相同。

下面介绍 INotifyPropertyChanged 接口功能,部分功能将在 5.3.4 节的案例中体现。

(1) 有两个集合类型,分别是 System.Collections.ObjectModel.ObservableCollection<T>和 System.ComponentModel.BindingList<T>。

(2) 有两个数据源,分别是 CompositeCollection 和 CollectionViewSource。

(3) 具有两个数据提供者,分别是 DataSourceProvider 和 XmlDataProvider。该类的派生类有 ObjectDataProvider。XmlDataProvider 允许用户访问 XML 数据。ObjectDataProvider 能够在 XAML 中以如下方式创建绑定源对象:使用 MethodName 和 MethodParameters 属性执行函数调用;使用 ObjectType 指定类型并通过 ConstructorParameters 属性将参数传递给对象的构造函数;直接为 ObjectInstance 属性赋值指定需要用作绑定源的对象。

5.3.4 绑定到列表框

在程序开发中,数据通常以列表的方式呈现。将数据绑定到 ListBox 是常见的数据绑定形式。现在要求实现如图 5.22 所示的页面效果。页面分为上、中、下三部分,上部分是用

ListBox呈现姓名；中间部分是两个TextBox,用于显示姓名与年龄；下面是4个按钮,分别用于显示第一条数据、前一条数据、下一条数据和最后一条数据。页面执行的业务逻辑是：当单击ListBox中的姓名时,两个TextBox中的值也随之改变,单击某按钮,页面上显示其相应的数据内容。

先创建User类,其数据结构与上节完全相同,略去类定义相同代码。再创建Users类,因为Users是作为User的数据集合使用,所以需要继承自ObservableCollection < User > ,为XAML代码中Windows中的资源做准备。其CS代码中仅一条语句如下。

```
class Users : ObservableCollection < User > { }
```

Users类继承ObservableCollection < User > ,是INotifyPropertyChanged接口的集合类型,该属性变更接口还有一个集合类型,是System.ComponentModel.BindingList < T > ,也很常用。接下来,编写前台设计页面相应的XAML代码如下。

```
<! -- 保留Window代码部分 -->
  xmlns:local = "clr - namespace:DataBindingListBox"
  Title = "MainWindow" Height = "228" Width = "283">
    < Window.Resources >
        < local:Users x:Key = "Userset">
            < local:User Name = "赵子龙" Age = "120 "/>
            < local:User Name = "谢浩然" Age = "10 "/>
            < local:User Name = "林黛玉" Age = "20 "/>
            < local:User Name = "花木兰" Age = "90 "/>
            < local:User Name = "诸葛亮" Age = "120 "/>
            < local:User Name = "穆桂英" Age = "90 "/>
            < local:User Name = "贾宝玉" Age = "20 "/>
            < local:User Name = "李婉童" Age = "10 "/>
        </local:Users >
    </Window.Resources >
    < Grid Name = "grid" Height = "195" DataContext = "{StaticResource Userset}">
        < TextBlock Height = "23" HorizontalAlignment = "Left" Margin = "6,71,0,0" Name =
            "textBlock1" Text = "Name:" VerticalAlignment = "Top" Width = "43" />
        < TextBlock Height = "23" HorizontalAlignment = "Left" Margin = "6,0,0,72" Name =
            "textBlock2" Text = "Age:" VerticalAlignment = "Bottom" Width = "43" />
        < TextBox Text = " { Binding Path = Name }" Name = " nameTextBox" Height = " 23"
            HorizontalAlignment = "Left" Margin = "129,68,0,0" VerticalAlignment = "Top"
            Width = "120"/>
        < TextBox Text = " { Binding Path = Age }" Name = " ageTextBox" Height = " 23"
            HorizontalAlignment = "Left" Margin = "129,100,0,0" VerticalAlignment = "Top"
            Width = "120"/>
        < TextBlock Height = "23" HorizontalAlignment = "Left" Margin = "6,10,0,0" Width = "58"
            Text = "DataSet:" Name = "textBlock3" VerticalAlignment = "Top"/>
        < ListBox ItemsSource = "{Binding}" IsSynchronizedWithCurrentItem = "True"
            SelectedValuePath = "Age" DisplayMemberPath = "Name" Height = "52" HorizontalAlignment =
            "Left" Margin = "70,10,0,0" Name = "userListBox" VerticalAlignment = "Top" Width =
            "179" >
        </ListBox >
        < Button Content = "First" Height = "23" HorizontalAlignment = "Left" Margin = "6,127,
            0,0" Name = "firstButton" VerticalAlignment = "Top" Width = "47" Click =
            "firstButton_Click" />
        < Button Content = "Previo" Height = "23" HorizontalAlignment = "Left" Margin = "70,
            127,0,0" Name = "previousButton" VerticalAlignment = "Top" Width = "47" Click =
            "previousButton_Click" />
```

```
            <Button Content = "Next" Height = "23" HorizontalAlignment = "Left" Margin = "137,127,
                    0,0" Name = "nextButton" VerticalAlignment = "Top" Width = "47" Click =
                    "nextButton_Click" />
            <Button Content = "Last" Height = "23" HorizontalAlignment = "Left" Margin = "202,127,
                    0,0" Name = "lastButton" VerticalAlignment = "Top" Width = "47" Click =
                    "lastButton_Click" />
    </Grid>
</Window>
```

分析上述 XAML 代码,解释新增语句及属性的作用。

(1) 在 Windows.Resources 调用 Users 类,并将其命名为 Userset,语句如下所示。

`<local:Users x:Key = "Userset">`

(2) Grid 中的数据上下文是调用在 Windows.Resources 中已定义好的静态资源 Userset,定义语句是"DataContext＝"{StaticResource Userset}""。

(3) 两个文本框的 Text 属性分别绑定 User 类中的 Name 和 Age 属性。语句如下。

`<TextBox Text = "{Binding Path = Name}" Name = "nameTextBox" >; <TextBox Text = "{Binding Path = Age}" Name = "ageTextBox">`

(4) 列表框中,数据源同步当前数据项、选择值的路径、显示成员路径分别对应的属性 ItemsSource、IsSynchronizedWithCurrentItem、SelectedValuePath 和 DisplayMemberPath。

下面在后台定义 GetUsersetView()方法,实现在列表框中加载数据功能,CS 代码如下。

```
using System.ComponentModel;
public partial class MainWindow : Window
{
    ICollectionView GetUsersetView()
    {   Users users = (Users)this.FindResource("Userset");
        return CollectionViewSource.GetDefaultView(users);
    }
    User user = new User();
    List<User> list = new List<User>();
    public MainWindow()
    {     InitializeComponent(); }
}
```

上面代码中的 GetUsersetView()方法的返回值类型是 ICollectionView,该集合类型具有管理当前数据集(自定义排序、筛选、分组等)功能。在 return 返回值中出现的关键字 CollectionViewSource 是数据源。

继续添加 4 个按钮(firstButton、previousButton、nextButton、lastButton)的双击事件,代码如下。

```
private void firstButton_Click(object sender, RoutedEventArgs e)
{   if (userListBox.SelectedItem!= null)
    { userListBox.SelectedIndex = 0; }
}
private void previousButton_Click(object sender, RoutedEventArgs e)
{   ICollectionView view = GetUsersetView();
    view.MoveCurrentToPrevious();
    if(view.IsCurrentBeforeFirst)
```

```
        { view.MoveCurrentToFirst(); }
    }
private void nextButton_Click(object sender, RoutedEventArgs e)
    {   ICollectionView view = GetUsersetView();
        view.MoveCurrentToNext();
        if (view.IsCurrentAfterLast)
        { view.MoveCurrentToLast(); }
    }
private void lastButton_Click(object sender, RoutedEventArgs e)
    {   if (userListBox.SelectedItem!= null)
        { userListBox.SelectedIndex = userListBox.Items.Count – 1; }
    }
```

运行完整的代码,页面显示效果如图 5.22 所示。单击 Next 按钮,页面运行效果如图 5.23 所示,TextBox 中的数据与 ListBox 中的数据项始终保持一致。

图 5.22　绑定 ListBox 对象

图 5.23　向前移动功能显示页面

5.4　小　　结

本章介绍了数据驱动模型、数据绑定原理及数据绑定用法。WPF 数据驱动 UI 的设计思想让数据成为应用程序的中心,真正体现了"程序＝数据＋算法"的本质。

习题与实验 5

1. 简述数据绑定的原理及目的。
2. 实现数据绑定多种控件组合,如图 5.24 所示。

图 5.24　绑定 ListBox 对象

3. 设计数据绑定到 ListBox。功能：选中左边列表框中的朝代，右边 ListBox 中显示该朝代的名人，如图 5.25 所示。例如，选中元朝，对应着元朝的名人。

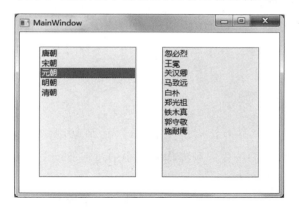

图 5.25　朝代与人物绑定页面

4. 在本章绑定到列表框的页面下部加入 Add、Sort、Filter、Group 这 4 个按钮，实现数据添加、按年龄排序、过滤、按年龄分组功能。程序运行后的初始页面如图 5.26 所示。单击 Group 按钮，列表框中的数据按年龄分组，页面显示效果如图 5.27 所示。

图 5.26　绑定到 ListBox 新添功能　　　　图 5.27　年龄分组页面

第 6 章

路由事件

Windows在操作系统平台占有绝对统治地位,基于Windows的编程和开发越来越广泛。学习本章路由事件时,需首先回顾操作系统中的过程驱动、消息机制、事件驱动模型,然后再学习WPF中的路由事件机制。

在早期的DOS(Disk Operating System,磁盘操作系统)下的任何程序都使用顺序的、过程驱动的程序设计方法。这种程序都有一个明显的开始、明显的过程及一个明显的结束,因此通过程序就能直接控制程序事件或过程的全部顺序。即使是在处理异常时,处理过程也仍然是顺序的、过程驱动的结构,所以DOS是过程驱动机制。而Windows是事件驱动的,事件驱动围绕着消息的产生与处理展开,事件驱动是靠消息循环机制来实现的。而WPF的路由事件,又扩充了事件驱动模型,通过自定义路由事件,能够实现用户更大的需求,以解决实际问题。下面从消息机制开启本章内容。

6.1 消息机制

在Windows中发生的一切都可以用消息来表示,消息用于告诉操作系统发生了什么,所有的Windows应用程序都是消息驱动的。一个消息是由消息的名称(UINT)和两个参数(WPARAM和LPARAM)组成的。消息的参数中包含有重要的信息。例如,对鼠标消息而言,LPARAM中一般包含鼠标的位置信息,而WPARAM参数中包含了发生该消息时,Shift、Ctrl等键的状态信息。对于不同的消息类型来说,两个参数也都相应地具有明确的意义。

6.1.1 消息的运行机制

消息常被说成Windows操作系统的灵魂,是掌握Windows编程的神兵利器。在路由事件学习前,先来了解一下消息运行机制。

1. 消息的概念

消息(Message)是指Windows操作系统发给应用程序的一个通告,它告诉应用程序某个特定的事件发生了。例如,用户单击鼠标或按键都会引发Windows系统发送相应的消息。最终处理消息的是应用程序的窗口函数,如果程序不处理,操作系统将会做出默认处理。消息包含了消息的类型标识符和其他附加信息。从数据结构的角度来说,它是一个结构体类型,系统定义的结构体MSG用于表示消息,定义形式如下:

```
typedef struct tagMSG
{   HWND hwnd;
    UINT message;
    WPARAM wParam;
    LPARAM lParam;
    DWORD time;
    POINT pt;
}MSG;
```

其中,hwnd 是窗口的句柄,这个参数将决定由哪个窗口过程函数对消息进行处理;message 是一个消息常量,用来表示消息的类型;wParam 和 lParam 都是 32 位的附加信息,具体表示什么内容,要视消息的类型而定;time 是消息发送的时间;pt 是消息发送时鼠标所在的位置。

2. Windows 编程原理

Windows 编程是通过消息机制来实现,故把 Windows 视为消息(Message)驱动式系统。消息机制实现了应用程序之间、应用程序与操作系统之间的通信方式。消息触发、消息响应及处理行为实现应用程序的功能。在 Windows 系统中,有系统消息队列和应用程序消息队列两种。Windows 监控所有输入设备。当一个事件被触发,Windows 先将输入的消息放入系统消息队列中,随后再将输入的消息复制到相应的应用程序队列中,应用程序中的消息循环从它的消息队列中检索每一个消息并发送给相应的窗口函数中。消息就是描述事件发生的信息,Windows 程序是事件驱动的,故此 Windows 程序的执行顺序由事件的发生顺序来决定,具有不可预知性。应用程序、操作系统和计算机硬件(输入输出设置)之间的关系如图 6.1 所示。

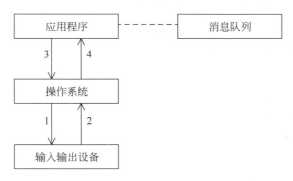

图 6.1 应用程序、操作系统和计算机硬件之间的关系

箭头 1 说明操作系统能够操纵输入输出设备,如让打印机打印。箭头 2 说明操作系统能够感知输入输出设备的状态变化,如鼠标单击、按键按下等,这就是操作系统和计算机硬件之间的交互关系。应用程序开发者并不需要知道它们之间是如何做到的,读者需要了解的是操作系统与应用程序之间如何交互。箭头 3 是应用程序通知操作系统执行某个具体的操作,这是通过调用操作系统的 API 来实现的;操作系统也能够感知硬件的状态变化,但并不决定如何处理,而是把这种变化转交给应用程序,由应用程序决定如何处理。向上的箭头 4 说明了这种转交情况,操作系统通过把每个事件都包装成一个称为消息结构体 MSG 来实

现这个过程，也就是消息响应。在掌握消息概念和 Windows 编程原理的前提下，再来认识消息循环。

3. Windows 消息循环

应用程序通过消息循环来获取各种消息，通过相应的窗口过程函数，对消息处理；消息循环让应用程序响应外部各种事件，消息循环是一个 Windows 应用程序的核心部分。

Windows 操作系统为每个线程维持一个消息队列，当事件产生时，操作系统感知这一事件的发生，并包装成消息发送到消息队列，应用程序通过 GetMessage() 函数取得消息并存于一个消息结构体中，通过 TranslateMessage() 解释消息，通过 Dispatch Message() 分发消息。下面的代码描述了 Windows 的消息循环。

```
while(GetMessage(&msg, NULL, 0, 0))
{   TranslateMessage(&msg) ;
    DispatchMessage(&msg) ;}
```

TranslateMessage(&msg) 用于解释键盘按键按下和弹起（分别对于 KeyDown 和 KeyUp 消息），产生一个 WM_CHAR 消息。对于大多数消息是不起作用的。

DispatchMessage(&msg) 把消息发到消息结构体中对应的窗口，由窗口过程函数处理消息。GetMessage() 在取得 WM_QUIT 之前返回值均为 TRUE，也就是说只有获取到 WM_QUIT 消息，才返回 FALSE，才能跳出消息循环。

4. 消息处理

消息由窗口处理函数进行处理。对于每个窗口类，Windows 都预备了一个默认的窗口过程处理函数 DefWindowProc()。这样做的好处是，可以着眼于用户感兴趣的消息，把其他不感兴趣的消息传递给默认窗口过程函数进行处理。每一个窗口类都有一个窗口过程函数，此函数是一个回调函数，它是由 Windows 操作系统负责调用的，而应用程序本身不能调用它。以 switch 语句开始，对于每条感兴趣的消息都以一个 case 引出。下面的代码描述了 Windows 的消息处理过程。

```
LRESULT CALLBACK WndProc(HWND hwnd, UINT message,WPARAM wParam,
        LPARAM lParam)
{
…
switch(uMsgId)
{case WM_TIMER:              //对 WM_TIMER 定时器消息的处理过程
return 0;
case WM_LBUTTONDOWN:         //对单击消息的处理过程
return 0;
…
default:
return DefWindowProc(hwnd,uMsgId,wParam,lParam);}
}
```

对于每条已经处理过的消息都必须返回 0；否则，消息将不停地重试下去；对于不感兴趣的消息，交给 DefWindowProc() 函数进行处理，并需要返回其处理值。消息处理机制如图 6.2 所示。

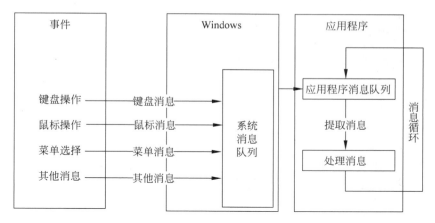

图 6.2 消息处理机制

6.1.2 事件模型

消息是事件的前身，消息的本质就是一组数据记录执行的操作，消息处理函数根据消息数据执行相应的操作。随着微软面向对象开发平台的日益成熟，消息机制被封装成事件模型。事件模型隐藏了消息机制的许多细节。当对象有相关的事件发生时（如按下鼠标键），对象产生一条特定的标识事件发生的消息，消息被送入消息队列，或不进入队列而直接发送给处理对象，主程序负责组织消息队列，将消息发送给相应的处理程序，使相应的处理程序执行相应的动作，做完相应的处理后将控制权交还给主程序，这就是事件驱动机制。Windows 就采用这种机制，程序的设计围绕事件模型来进行。在这种机制中，对象的请求仅仅是向队列中添加相应的消息，耗时的处理则被分离给处理函数。这种结构的程序中各功能模块界限分明，便于扩充，能充分地利用 CPU 的处理能力，使系统对外界响应准确而及时。

事件模型由事件拥有者、事件、事件的处理器及订阅关系这 4 个部分组成，以 button1 按钮对象为例，说明如下。

事件的拥有者就是 button1（按钮）；事件是 button1.Click；事件的处理器，在 CS 代码中是 button1_Click。其中，订阅关系是通过一条语句，让事件和事件处理器建立联系。语句是"this.button1.Click＋＝new System.EventHandler(this.button1_Click);"，当然，一个事件也可以定义多个处理器响应。最后一个响应者，就是窗体本身。

对事件模型的理解，还有一种提法是事件模型三要素。三要素分别是指事件源、监听器和事件处理程序。事件源能够接收外部事件的源体；监听器能够接收事件源通知的对象；事件处理程序用于处理事件的对象函数。根据事件驱动模型三要素，事件模型如图 6.3 所示。

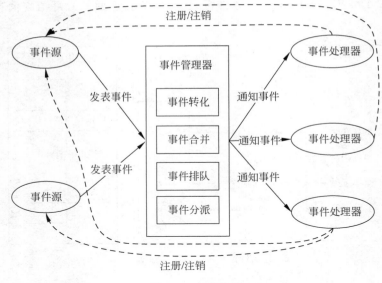

图 6.3 事件模型

6.2 路由事件原理

传统的事件模型中，在消息激发时，消息被订阅后，交给事件的响应者，事件的响应者使用事件的处理器来做出响应。既然.NET 中已经有事件机制了，为什么在 WPF 中引入路由事件来取代事件呢？因为 WPF 采用元素合成的设计思想，WPF 有 LogicalTree 和 VisualTree 两棵树，WPF 中界面元素可能是由多个元素合成而来的，以一个 Button 上放置一张图片为例，单击图片与单击 Button，外界如何辨别是谁被单击了？这就需要用路由事件。

在过去的事件模型中，事件源（事件的宿主）必须能够直接访问到事件的响应者，它们是直接的订阅关系。在 WPF 中的路由事件中，事件源与事件的响应者之间则没有直接的显式订阅关系。事件的拥有者则只负责激发事件，事件将由谁响应，它并不知道，事件的响应者则通过事件的监听器，当有此类事件传递过来后，事件响应者就使用事件处理器来响应事件，并决定此事件是否继续传递。例如，一个 Button 被"单击"后，事件被触发。然后事件就会沿着逻辑树进行传递，事件的响应者安装了监听器，当监听到这个事件进行响应，并决定这个事件是否继续传递。若事件在某个结点处理以后，不想让它继续传递，可以把它标记为"已处理"（把 Handled 属性设为 true），则路由事件停止传递。因为所有的路由事件都共享一个公共的、事件数据基类 RoutedEventArgs。RoutedEventArgs 定义了一个采用布尔值的 Handled 属性。

6.2.1 路由事件机制

以鼠标右击传递路由事件为例，设计从外向内依次为黄、红、粉 3 个 Grid 嵌套，中心处放置一个 Button 按钮，XAML 页面显示效果如图 6.4 所示。

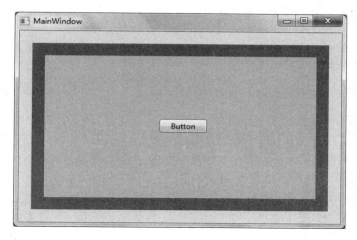

图 6.4 路由事件机制

该页面的 XAML 代码如下。

```xml
<!-- 保留 Window 代码部分 -->
    Title = "MainWindow" Height = "350" Width = "525" MouseUp = "Window_MouseUp" >
    <Grid>
        <Grid Background = "Yellow" MouseUp = "Grid_MouseUp" Name = "gridYellow">
            <Grid Margin = "20" Background = "red" MouseUp = "Grid_MouseUp" Name = "gridRed">
                <Grid Margin = "20" Background = "Pink" MouseUp = "Grid_MouseUp" Name = "gridPink">
                    <Button x:Name = "button" Content = "Button" HorizontalAlignment = "Left"
                            Margin = "182,103,0,0" VerticalAlignment = "Top" Width = "75"
                            Click = "button_Click" MouseUp = "button_MouseUp" />
                </Grid>
            </Grid>
        </Grid>
    </Grid>
</Window>
```

在后台用 CS 编写相应事件的代码如下。

```csharp
private void Window_MouseUp(object sender, MouseButtonEventArgs e)
{
    MessageBox.Show("Window 被点击");
    MessageBox.Show("路由事件结束");
}
private void Grid_MouseUp(object sender, MouseButtonEventArgs e)
{
    Grid g = sender as Grid;
    MessageBox.Show(g.Name + "被点击");
}
private void button_MouseUp(object sender, MouseButtonEventArgs e)
{
    FrameworkElement f = sender as FrameworkElement;
```

```
        MessageBox.Show(f.Name + "被点击");
        //e.Handled = true;
}
private void button_Click(object sender, RoutedEventArgs e)
{
        MessageBox.Show(((Button)sender).Name);
}
```

运行上述代码,显示初始页面如图 6.4 所示。当单击 Button 按钮时,弹出消息框,如图 6.5 所示。单击"确定"按钮后,button_Click 事件终止。当右击 Button 按钮,触发 button_MouseUp 事件,弹出消息框如图 6.6 所示。单击图 6.6 中的"确定"按钮后,弹出如图 6.7 所示的 gridPink 消息框;单击图 6.7 中的"确定"按钮后,弹出如图 6.8 所示的 gridRed 消息框;单击图 6.8 中的"确定"按钮后,弹出如图 6.9 所示的 gridYellow 消息框;单击图 6.9 中的"确定"按钮后,又弹出如

图 6.5 单击 Button 消息框

图 6.10 所示的 Window 消息框;单击图 6.10 中的"确定"按钮后,弹出如图 6.11 所示的路由事件结束消息框。

图 6.6 右击 Button 消息框

图 6.7 gridPink 消息框

图 6.8 gridRed 消息框

图 6.9 gridYellow 消息框

图 6.10 Window 消息框

图 6.11 路由事件结束消息框

在 gridPink 处单击,会依次弹出如图 6.8～图 6.11 所示的消息框;在 gridRed 处单击,会依次弹出如图 6.9～图 6.11 所示的消息框;在 gridYellow 处单击,会依次弹出如图 6.10～图 6.11 所示的消息框。

让读者疑惑的是,单击的是按钮,为什么所有的 Grid 和 Window 也会引发事件呢?其实这就是 WPF 路由事件的机制,引发的事件由源元素逐级传到上层的元素,路由事件会沿着逻辑树:button→gridPink→gridRed→gridYellow→Window。当右击 button 时,触发 button 的 Mouseup 路由事件,该路由事件沿着逻辑树向上传播,触发 Grid 的 MouseUp 路

由事件,继续向上,直到最外层 Window 的 MouseUp 后,路由事件终止。

若想在图 6.7 消息框弹出后就终止路由,则需要修改在 Grid_MouseUp 中的 Handled 属性值为 true。后台 CS 代码中修改如下。

```
private void Grid_MouseUp(object sender, MouseButtonEventArgs e)
{
    Grid g = sender as Grid;
    MessageBox.Show(g.Name + "被点击");
    e.Handled = true;
}
```

在第 1 章中就提及了 WPF 的逻辑树,接下来通过逻辑树来深入理解 WPF 路由事件调用机制。设计页面如图 6.12 所示,其 XAML 代码如下所示。图 6.12 路由事件页面的 LogicalTree 结构如图 6.13 所示。

图 6.12 路由事件页面

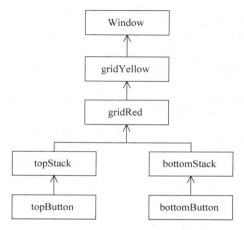

图 6.13 路由事件页面的 LogicalTree 结构

```
<!-- 保留 Window 代码部分 -->
<Grid>
    <Grid x:Name = "gridYellow" Background = "Yellow">
        <Grid x:Name = "gridRed" Margin = "10" Background = "Red">
            <Grid.RowDefinitions>
                <RowDefinition/>
                <RowDefinition/>
            </Grid.RowDefinitions>
            <StackPanel x:Name = "topStack" Grid.Row = "0" Background = "Pink" Margin = 
             "10">
                <Button x:Name = "topButton" Content = "Top" Width = "90" Height = "100"
                 Margin = "15"/>
            </StackPanel>
            <StackPanel x:Name = "bottomStack" Grid.Row = "1" Background = "Pink" Margin
             = "10">
                <Button x:Name = "bottomButton" Content = "Bottom" Width = "90" Height = 
                 "100" Margin = "15" MouseUp = "bottomButton_MouseUp"/>
```

```
            </StackPanel>
        </Grid>
      </Grid>
    </Grid>
</Window>
```

在后台编写 Grid_MouseUp、bottomStack_MouseUP 和 bottomButton_MouseUp 右击路由事件，CS 代码如下。

```
private void Grid_MouseUp(object sender, MouseButtonEventArgs e)
{
    Grid g = sender as Grid;
    MessageBox.Show(g.Name + "被点击");
}
private void bottomStack_MouseUP(object sender, MouseButtonEventArgs e)
{
    StackPanel s = sender as StackPanel;
    MessageBox.Show(s.Name + "被点击");
}
private void bottomButton_MouseUp(object sender, MouseButtonEventArgs e)
{
    MessageBox.Show(((Button)sender).Name + "被点击");
}
```

运行上述代码，右击 bottomButton 按钮，没有响应。前个案例中，是在前台 XAML 页面编写了路由事件。实现右击后，激活路由有关 MouseUp 路由事件，实现沿逻辑树向上传播。接下来，在后台窗体构造器中添加的 CS 代码如下。

```
public MainWindow()
{
    InitializeComponent();
    gridYellow.AddHandler(UIElement.MouseUpEvent,
        new MouseButtonEventHandler(Grid_MouseUp));
    gridRed.AddHandler(UIElement.MouseUpEvent,
        new MouseButtonEventHandler(Grid_MouseUp));
    bottomStack.AddHandler(UIElement.MouseUpEvent, new
        MouseButtonEventHandler(bottomStack_MouseUP));
}
```

运行上述代码，右击 bottomButton 按钮，依次弹出如图 6.14～图 6.17 所示的消息框。

图 6.14　bottomButton 消息框

图 6.15　bottomStack 消息框

图 6.16　gridRed 消息框

图 6.17　gridYellow 消息框

读者看到,调用 UIElement. AddHandler()方法是为了把监听的事件与事件处理器关联起来。AddHandler()方法源自 UIElement 类,该方法的第一个参数是 UIElement. MouseUpEvent,这就意味着,路由事件本身是一个 RoutedEvent 类型的静态成员变量。

在用 AddHandler()方法创建委托类型(如 MouseButtonEventHandler),不能隐式地创建委托对象。原因是 UIElement. AddHandler()方法支持所有的 WPF 事件,它并不知晓用户想要使用的委托类型。

6.2.2　RoutedEventArgs 类

上述路由事件中,sender 参数为路由事件的传播提供引用,在有些情况下,需要确定事件最初发生的位置。可以通过 RoutedEventArgs 类的属性得到细节信息。因为所有的 WPF 事件参数类继承自 RoutedEventArgs,任何事件处理程序都可以使用这些属性,表 6.1 列出了 RoutedEventArgs 的属性含义。

表 6.1　RoutedEventArgs 类的属性

属　性　名	含　　义
Source	触发事件的对象
OriginalSource	引发事件的初始对象。通常其与 Source 的属性相同。但在某些情况下,OriginalSource 属性指向更深的层次,以获得作为更高一级元素一部分的后台元素
RoutedEvent	通过事件处理程序为触发事件提供 RoutedEvent 对象(如静态的 UIElement. MouseUpEvent)。当使用同一事件处理程序处理不同的事件时,该信息极为重要
Handled	用于终止路由事件。Handled 属性为 true 时,路由事件终止

传统的. NET Event 机制就是单纯的委托,而 RoutedEvent 则提供了在 Logical Tree 中路由事件的能力。

6.2.3　路由策略

到目前为止,所有的路由事件都是冒泡路由事件,WPF 中的路由策略除了冒泡,还有隧道和直接模式。

(1) 冒泡(Bubbling):由事件源沿着逻辑树向上传递,直到根元素。

(2) 隧道(Tunneling):与冒泡相反,从元素树的根部调用事件处理程序并依次向下深入直到事件源。一般情况下,WPF 提供的输入事件都是以隧道/冒泡对实现的。例如,PreviewKeyDown 事件将首先从顶级窗体开始触发,直到获得用户焦点所在的元素上。隧道事件的命名要求以 Preview 开头,故常被称为 Preview 事件。

(3) 直接(Direct):只有事件源才有机会响应事件,就是 CLR 事件。

现在对图 6.12 中的 topButton 按钮来编写隧道模式的路由事件。先在后台编写 topButton_PreviewMouseDown 事件,该事件的 CS 代码如下。

```
private void topButton_PreviewMouseDown(object sender, MouseButtonEventArgs e)
{
    MessageBox.Show(sender.GetType().ToString());
}
```

接下来写前台的 XAML 代码,因后台已编写 topButton_PreviewMouseDown 路由事件,故在前台添加 PreviewMouseDown,再编写 topButton_PreviewMouseDown 事件处理程序(建议根据系统的智能提示选择该事件的处理程序),XAML 代码如下。

```
< Window x:Class = "RoutedEventMechanism1.MainWindow"
        xmlns = "http://schemas.microsoft.com/winfx/2006/xaml/presentation"
        xmlns:x = "http://schemas.microsoft.com/winfx/2006/xaml"
        Title = "MainWindow" Height = "350" Width = "525"
        PreviewMouseDown = "topButton_PreviewMouseDown" >
    < Grid >
        < Grid x:Name = "gridYellow" Background = "Yellow"
              PreviewMouseDown = "topButton_PreviewMouseDown">
            < Grid x:Name = "gridRed" Margin = "10" Background = "Red" >
                < Grid.RowDefinitions >
                    < RowDefinition/>
                    < RowDefinition/>
                </Grid.RowDefinitions >
                < StackPanel x:Name = "topStack" Grid.Row = "0" Background = "pink"
                    Margin = "10" PreviewMouseDown = "topButton_PreviewMouseDown" >
                < Button x:Name = "topButton" Content = "Top" Width = "90" Height = "100"
                    Margin = "15" PreviewMouseDown = "topButton_PreviewMouseDown" />
                </StackPanel >
                < StackPanel x:Name = "bottomStack" Grid.Row = "1" Background = "Pink"
                        Margin = "10" >
                    < Button x:Name = "bottomButton" Content = "Bottom" Width = "90"
                                    Height = "100" Margin = "15" />
                </StackPanel >
            </Grid >
        </Grid >
    </Grid >
</Window >
```

运行上述代码,右击 topButton 按钮后,依次弹出如图 6.18~图 6.21 所示的消息框。

图 6.18 MainWindow 消息框

图 6.19 Grid 消息框

图 6.20　StackPanel 消息框

图 6.21　Button 消息框

隧道模式路由事件会沿着逻辑树从 MainWindow→Grid→StackPanel→Button。当右击 topButton 时，触发 topButton_PreviewMouseDown 的隧道路由事件。隧道模式路由事件沿着逻辑树从最外层的 MainWindow 开始，向下传播，直到最内层的 topButton 后，路由事件终止。

直接模式路由事件就是 CLR 事件，在此不再举例说明。隧道模式和冒泡模式事件对比，如图 6.22 所示。

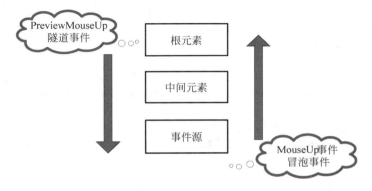

图 6.22　隧道模式事件与冒泡模式事件对比

6.3　自定义路由事件

为了便于程序中对象间通信，用户常常需要自定义一些路由事件。自定义路由事件可分为下列 3 个步骤。

（1）声明并注册路由事件。
（2）为路由事件添加 CLR 事件包装。
（3）创建激发路由事件的方法。

下面自定义一个 WPF 路由事件，该事件是冒泡式路由事件，其功能是：单击页面中的按钮后，给出当前时间及冒泡路由的控件。该路由事件具有两个参数，先来创建类，让其继承自 RoutedEventArgs 类，该类名为 RecordingTimeEventArgs。该类的 CS 代码如下。

```
class RecordingTimeEventArgs:RoutedEventArgs
{
    public RecordingTimeEventArgs（RoutedEvent routedEvent, object source）: base
```

```
(routedEvent, source) { }
        public DateTime ClickTime { get; set; }
}
```

再创建一个类,类名为 TimerButton,其继承自 Button 类。在该类中实现自定义路由事件时,遵照上述所说的"自定义路由事件的 3 个步骤"。

```
class TimerButton:Button
{
        //声明注册路由事件
        public static readonly RoutedEvent RecordingTimeEvent = EventManager.RegisterRoutedEvent
        ("RecordingTime",RoutingStrategy.Bubble,typeof(EventHandler < RecordingTimeEventArgs >),
        typeof(TimerButton));
        //CLR 事件包装
        public event RoutedEventHandler RecordingTime
        {
            add { this.AddHandler(RecordingTimeEvent,value); }
            remove { this.RemoveHandler(RecordingTimeEvent,value ); }
        }
        //激发路由事件,借用 Click 事件的激发方法
        protected override void OnClick()
        {
            base.OnClick();      //保证 Button 原有功能正常使用,Click 事件被激发
            RecordingTimeEventArgs args = new RecordingTimeEventArgs(RecordingTimeEvent,
            this);
            args.ClickTime = DateTime.Now;
            this.RaiseEvent(args); //UIElement 及其派生类
        }
}
```

对上述声明注册路由事件的 4 个参数说明如下。

第 1 个参数:RecordingTime 为路由事件的名称。

第 2 个参数:RoutingStrategy.Bubble 是路由事件的冒泡策略,共有 3 种策略分别是 Bubble 冒泡模式、Tunnel 隧道模式和 Direct 直接模式。

第 3 个参数:typeof(EventHandler < RecordingTimeEventArgs >)指定事件处理器的类型。

第 4 个参数:typeof(TimerButton)指定事件的宿主类型。

现在开始布局前台页面,页面的效果如图 6.23 所示,其前台页面的 XAML 如下。

```
< Window x:Class = "CustomRoutedEvent.MainWindow"
        xmlns = "http://schemas.microsoft.com/winfx/2006/xaml/presentation"
        xmlns:x = "http://schemas.microsoft.com/winfx/2006/xaml"
        xmlns:local = "clr - namespace:CustomRoutedEvent"
        local:TimerButton.RecordingTime = "RecordingTimeHandler"
        Title = "MainWindow" Height = "350" Width = "525">
    < Grid x:Name = "yellowGrid" Background = "Yellow"
        local:TimerButton.RecordingTime = "RecordingTimeHandler">
        < Grid x:Name = "pinkGrid" Background = "Pink" Margin = "20"
```

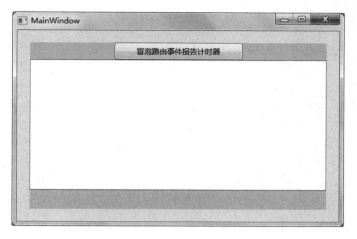

图 6.23 自定义路由事件

```
            local:TimerButton.RecordingTime = "RecordingTimeHandler">
        < StackPanel x:Name = "stackPanel1"
                local:TimerButton.RecordingTime = "RecordingTimeHandler">
        <local:TimerButton x:Name = "timeButton" Content = "冒泡路由事件报告计时器"
            local:TimerButton.RecordingTime = "RecordingTimeHandler"
                Height = "28" Width = "202" />
            < ListBox x:Name = "timeListBox" Height = "213" />
        </StackPanel>
        </Grid>
    </Grid>
</Window>
```

后台编写 RecordingTimeHandler 方法,该方法的 CS 代码如下。

```
public MainWindow()
{
    InitializeComponent();
}
private void RecordingTimeHandler(object sender, RecordingTimeEventArgs e)
{
    FrameworkElement element = (FrameworkElement)sender;
    string timeStr = e.ClickTime.ToString("HH:mm:ss");
    string content = string.Format("{0}到达{1}的时间: ", timeStr, element.Name);
    this.timeListBox.Items.Add(content);
}
```

运行完整的代码,在单击图 6.23 中的按钮后,页面运行效果如图 6.24 所示。在时间列表框(timeListBox)中显示自定义冒泡路由事件时间顺序:timeButton→stackPanel1→pinkGrid→yellowGrid。从里向外,并记录冒泡的时间。再次单击"冒泡路由事件报告计时器"按钮,在 timeListBox 列表框中继续显示本次的冒泡过程及时间。

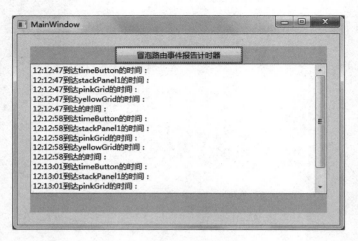

图 6.24 自定义路由事件响应 Button 事件

6.4 附加事件

WPF 中还有一种——附加事件，正如前面说的路由事件的宿主都是可以看到的界面元素，但是附加事件不具备在用户界面上显示的能力，如一个文本框的改变、鼠标的按下、键盘的按下。当然，"附加事件"也是路由事件，常见类及其对应的附加事件如下。

- Binding 类：SourceUpdated 事件、TargetUpdated 事件；
- Mouse 类：MouseEnter 事件、MouseLeave 事件、MouseDown 事件、MouseUp 事件等；
- Keyboard 类：KeyDown 事件、KeyUp 事件等。

6.5 小 结

本章从 Windows 操作系统的消息机制出发，介绍了事件模型，由于 WPF 的合成特性、内容模型，其引入路由事件机制，并采用冒泡、隧道、直接 3 种策略。用户可以通过自定义路由事件，写出紧凑、组织良好的代码，实现用户需求。

习题与实验 6

1. 简述消息处理机制、事件模型及数据驱动 UI，寻找它们在生活中的应用实例。
2. 对比分析路由事件与普通 CLR 事件的区别。
3. 路由事件的优点是什么？
4. 实现自定义路由事件，分别用隧道、冒泡和直接 3 种策略实现。隧道策略弹出页面的顺序如图 6.25 所示。冒泡策略的弹出页面的顺序如图 6.26 所示。

操作提示：文档结构如图 6.27 所示。其中，DetailReportButton 类由 Button 派生，它包含定义路由事件、CLR 事件包装和事件的触发方法。DetailReportEventArgs 是RoutedEventArgs 的派生类。前台 XAML 页面的 Logical Tree 如图 6.28 所示。

第 6 章 路由事件

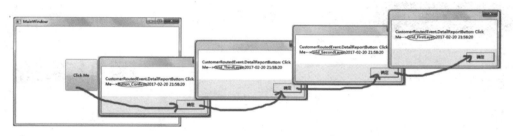

图 6.25 冒泡策略的弹出页面顺序

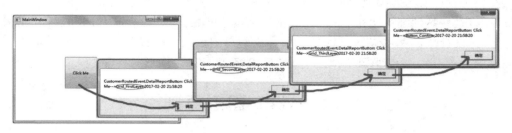

图 6.26 隧道策略的弹出页面顺序

图 6.27 本题文档结构

图 6.28 XAML 页面的 Logical Tree

由图 6.28 可知,MainWindow.xaml 的页面嵌套了 3 个 Grid,其 Grid 的名称从外向内,分别是 Grid_FirstLayer、Grid_SecondLayer 和 Grid_ThirdLayer。最内层 Grid 布局中还有一个 DetailReportButton,其名称是 Button_Confirm。

第 7 章

图形基础

在第 1 章读者已经知道 WPF 集成了 2D 矢量图、光栅图片、文本、音频、视频、动画、3D 图形，所有这些特性都构建在 DirectX 之上。从本章开始，读者需学习 WPF 的图形系统。

WPF 设计中的基本设计原则是元素合成的思想。元素合成的思想已融入 WPF 的图形系统中。本章从 WPF 图形原则开始，然后讲解 2D 和 3D 图形的要素。

7.1 WPF 图形原则

WPF 图形原则秉承元素合成的设计思想，WPF 的应用程序与分辨率无关（即应用程序不管设备的分辨率是多少，运行的效果都是一样的）。

那么 WPF 的应用程序为什么与分辨率无关呢？因为 WPF 的坐标并不是物理像素，而是逻辑像素。它使用的坐标系统等于 1/96 英寸。

WPF 的图形系统是基于变换 X 和 Y 坐标值来表示位置。常见的 3 个变换对元素（TranslateTransform、ScaleTransform 和 RotateTransform）进行定位、改变尺寸及旋转。

7.1.1 几何图形与笔刷

1. 几何图形

在 WPF 中，Geometry（几何图形）是所有 2D 图形的基础。PathGeometry（路径几何图形）又是所有其他几何图形的集合。故此先来了解 Path（路径）。

路径是由一系列的图案构成的。每个图案又是由一组线段构成的。WPF 中的线段有 LineSegment、BezierSegment 和 ArcSegment。其中，最简单的线段是 LineSegment。线段都需要有起点，当路径封闭时，设置"IsClosed="True""即可。

现在使用 Path 形状，定义封闭的路径。XAML 代码如下。

```xml
<Grid>
    <Path Fill = "Pink" Stroke = "Yellow"
        StrokeThickness = "4">
        <Path.Data>
            <PathGeometry>
                <PathGeometry.Figures>
                    <PathFigure StartPoint = "28,10" IsClosed = "True">
                        <LineSegment Point = "10,20"/>
                        <LineSegment Point = "20,40"/>
                        <LineSegment Point = "40,70"/>
                        <LineSegment Point = "70,80"/>
```

```xml
            <LineSegment Point = "80,120"/>
            <LineSegment Point = "120,10"/>
          </PathFigure>
        </PathGeometry.Figures>
      </PathGeometry>
    </Path.Data>
  </Path>
 </Grid>
</Window>
```

运行上述代码,页面显示效果如图 7.1 所示。

下面来认识贝塞尔线段(BezierSegment)。使用它时,需要通过两个控制点(Point1 和 Point2)和一个结束点(Point3)来定义一个贝塞尔曲线(Beziercurve)。XAML 代码如下。

图 7.1 LineSegment 页面显示效果

```xml
<PathGeometry.Figures>
    <PathFigure StartPoint = "90,5" IsClosed = "True">
        <LineSegment Point = "10,10"/>
          <BezierSegment Point1 = "260,40" Point2 = "400,430" Point3 = "700,300" IsSmoothJoin = "True"/>
    </PathFigure>
</PathGeometry.Figures>
```

运行上述代码,页面显示效果如图 7.2 所示。

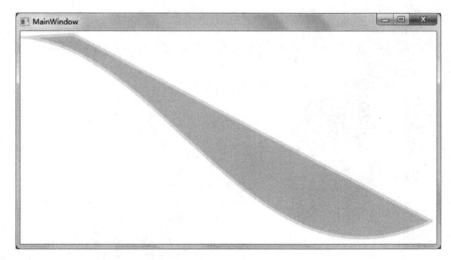

图 7.2 BezierSegment 页面显示效果

接下来再认识弧线段(ArcSegment)。弧线段是用于产生一个给定尺寸的椭圆。在贝塞尔线段(BezierSegment)的基础上实现椭圆效果,XAML 代码如下。

```xml
<PathFigure StartPoint = "90,5" IsClosed = "True">
    <LineSegment Point = "10,10"/>
    <BezierSegment Point1 = "260,40" Point2 = "400,430" Point3 = "700,300" IsSmoothJoin = "True"/>
    <ArcSegment Point = "70,100" Size = "65,55" IsLargeArc = "False" SweepDirection =
```

```
            "Clockwise" />
</PathFigure>
```

运行上述代码,页面显示效果如图 7.3 所示。若在</PathFigure>前加上 Quadratic-BezierSegment 的对象约束,语句内容"< QuadraticBezierSegment Point1＝"80,45" Point2＝"9,79"/>",再运行代码,页面显示效果如图 7.4 所示。

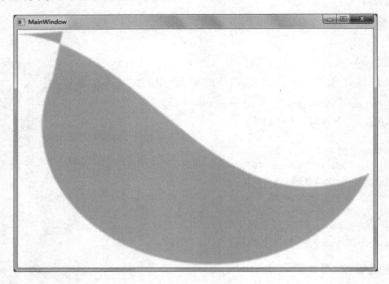

图 7.3　ArcSegment 页面显示效果

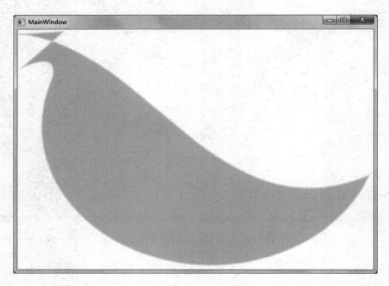

图 7.4　QuadraticBezierSegment 页面显示效果

几何图形中除了上述所说的各种线段外,还有 CombineGeometry 和 GeometryGroup 两个混合几何图形。

CombineGeometry 用来创建新的单独的几何图形,使用 GeometryCombineMode 决定

系统组合几何图形的模式,最终结果的外部轮廓会被勾勒出来。下面观察不同模式下的组合几何图形。

```
<Grid>
    <Path Stroke = "Yellow" StrokeThickness = "1" Fill = "Pink">
        <Path.Data>
            <CombinedGeometry GeometryCombineMode = "Exclude">
                <CombinedGeometry.Geometry1>
                    <EllipseGeometry RadiusX = "40" RadiusY = "40" Center = "75,75" />
                </CombinedGeometry.Geometry1>
                <CombinedGeometry.Geometry2>
                    <EllipseGeometry RadiusX = "40" RadiusY = "40" Center = "125,75" />
                </CombinedGeometry.Geometry2>
            </CombinedGeometry>
        </Path.Data>
    </Path>
</Grid>
</Window>
```

运行上述代码,页面显示效果如图 7.5 所示。当"GeometryCombineMode＝"Intersect""时,页面显示效果如图 7.6 所示；当"GeometryCombineMode＝"Union""时,页面显示效果如图 7.7 所示；当"GeometryCombineMode＝"Xor""时,页面显示效果如图 7.8 所示。

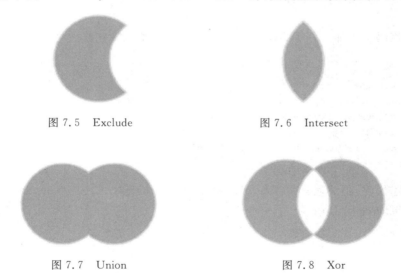

图 7.5　Exclude　　　　　　　图 7.6　Intersect

图 7.7　Union　　　　　　　　图 7.8　Xor

GeometryGroup(几何组合)继承自 Geometry 类。Geometry 类除含几何组合外,还包含几何线条(LineGeometry)、几何矩形(RectangleGeometry)、几何椭圆图形(EllipesGeometry)、几何路径(PathGeometry)。该类用来描述任何几何的 2D 形状。从绘图角度来看,Geometry 类和 Share 类都用来绘制 2D 图形,但是 Geometry(几何绘图)绘图效率高于 Share 类。

在 GeometryGroup 中创建 3 个 EllipseGeometry,填充规则是 EvenOdd,XAML 代码如下。

```
<Grid>
```

```
                < Path Stroke = "Yellow" StrokeThickness = "1" Fill = "Pink" >
                    < Path.Data >
                        < GeometryGroup FillRule = "EvenOdd" >
                            < EllipseGeometry Center = "40,70" RadiusX = "30" RadiusY = "30" />
                            < EllipseGeometry Center = "70,70" RadiusX = "30" RadiusY = "30" />
                            < EllipseGeometry Center = "50,40" RadiusX = "30" RadiusY = "30" />
                        </GeometryGroup>
                    </Path.Data>
                </Path>
        </Grid>
</Window>
```

运行上述代码,页面显示效果如图 7.9 所示。将填充规则换成 Nonzero 后,运行上述代码,页面显示效果如图 7.10 所示。

 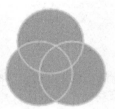

图 7.9　EvenOdd　　　　　　　　　图 7.10　Nonzero

在上述代码中再加入一个 RectangleGeometry,将填充规则修改为 EvenOdd,XAML 代码如下。

```
< GeometryGroup FillRule = "EvenOdd" >
    < EllipseGeometry Center = "40,70" RadiusX = "30" RadiusY = "30" />
    < EllipseGeometry Center = "70,70" RadiusX = "30" RadiusY = "30" />
    < EllipseGeometry Center = "50,40" RadiusX = "30" RadiusY = "30" />
    < RectangleGeometry Rect = "5,88 100 38" />
</GeometryGroup>
```

运行上述代码,页面显示效果如图 7.11 所示。将填充规则换成 Nonzero 后,运行代码,页面显示效果如图 7.12 所示。

图 7.11　组合 Nonzero　　　　　　　图 7.12　组合 EvenOdd

2. 笔刷

笔刷(Brush)是向系统发出在特定区域内绘制像素的命令。在有些示例中,这个区域是对路径的填充。WPF 有 6 种笔刷:SolidColorBursh、LinearGradientBrush、ImageBrush、

RadialGradientBrush、DrawingBrush、VisualBrush。下面分别用前 4 种笔刷对矩形框填充。4 种笔刷对应的效果如图 7.13～图 7.16 所示。

图 7.13 单色笔刷

图 7.14 线性渐变笔刷

图 7.15 径向渐变笔刷

图 7.16 图像笔刷

（1）SolidColorBursh（单色笔刷）的 XAML 代码如下。

```
<Grid>
    <Rectangle Width = "95" Height = "85">
        <Rectangle.Fill>
            <SolidColorBrush Color = "Orange" />
        </Rectangle.Fill>
    </Rectangle>
</Grid>
</Window>
```

（2）LinearGradientBrush（线性渐变笔刷）的 XAML 代码如下。

```
<Grid>
    <Rectangle Width = "95" Height = "85">
        <Rectangle.Fill>
            <LinearGradientBrush>
                <GradientStop Color = "Red" Offset = "0.0" />
                <GradientStop Color = "Orange" Offset = "0.6" />
                <GradientStop Color = "Yellow" Offset = "1.0" />
            </LinearGradientBrush>
        </Rectangle.Fill>
    </Rectangle>
</Grid>
</Window>
```

（3）RadialGradientBrush（径向渐变笔刷）的 XAML 代码如下。

```
<Rectangle Width = "95" Height = "85">
```

```xml
<Rectangle.Fill>
    <RadialGradientBrush GradientOrigin = "0.75,0.25">
        <GradientStop Color = "Red" Offset = "0.0" />
        <GradientStop Color = "Orange" Offset = "0.6" />
        <GradientStop Color = "Yellow" Offset = "1.0" />
    </RadialGradientBrush>
</Rectangle.Fill>
</Rectangle>
```

(4) ImageBrush(图像笔刷)的 XAML 代码如下。

```xml
<Rectangle Width = "75" Height = "75">
    <Rectangle.Fill>
        <ImageBrush ImageSource = "Images\flower.jpg"/>
    </Rectangle.Fill>
</Rectangle>
```

3. 画笔

画笔(Pen)用来勾勒出几何形状的轮廓,是由笔刷和笔的精细度组成的。在诸多示例中,容器元素的属性与画笔的属性是一致的。例如,画笔在矩形填充时,使用 StrokeThickness、StrokeDashArray 和 Stroke 等属性;在绘制边框时,使用 BorderBrush 和 BorderThickness 这两个属性。

7.1.2 绘制图画

在了解几何图形、笔刷及画笔的基础上,再来认识绘制图画(Drawing)。绘制图画通过自己的共享机制(Sharing)将显示特性变成为绘制图画对象,并赋予它一个矩形的几何形状。

到目前为止,章节中出现的概念都是围绕着树(Logical Tree 或 Visual Tree)来讲解的。但是绘制图画引入了图结构(Graph Structure)。图结构允许在图中多处显示单独的一幅图画,以提高性能。

下面演示共享模型。先定义一个 Ellipse 几何图形,使用 sharing 变量来共享。再定义两个 DrawingGroup 对象,把 sharing 作为它们的子对象。其中用到的 EllipseGeometry 对象和生成结果的 GeometryDrawing 都只被定义了一次。接下来,该图形可以在两个不同的上下文中共享。在此,每个 DrawingGroup 实例都使用了这个绘制图画对象,并在上面设置了 TranslateTransform。其中,XAML 代码只有一条语句 <Ellipse x:Name="ellipse"/>。后台 CS 代码如下。

```csharp
public MainWindow()
{
    InitializeComponent();
    GeometryDrawing sharing = new GeometryDrawing(Brushes.Green,
            null, new EllipseGeometry(new Rect(8, 8, 10, 10)));
    DrawingGroup a = new DrawingGroup();
    DrawingGroup b = new DrawingGroup();
    DrawingGroup c = new DrawingGroup();
    a.Children.Add(sharing);
```

```
            a.Transform = new TranslateTransform(0,0);
            b.Children.Add(sharing);
            b.Transform = new TranslateTransform(9,9);
            c.Children.Add(b);
            c.Children.Add(a);
            DrawingBrush brush = new DrawingBrush();
            brush.Drawing = c;
            brush.Viewport = new Rect(0,0,30,30);
            brush.ViewportUnits = BrushMappingMode.Absolute;
            brush.TileMode = TileMode.FlipXY;
            ellipse.Fill = brush;
        }
```

运行上述代码,页面显示效果如图 7.17 所示。

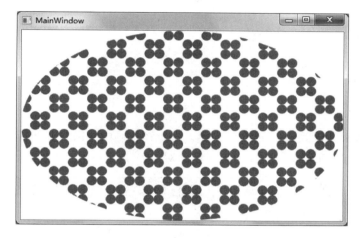

图 7.17　笔刷与几何图形绘制页面效果

绘制图画(Drawing)描述一些可见内容,如形状、位图、视频或一行文本。不同类型的绘图对象描绘的是不同类型的内容。WPF 具有 5 种不同类型对象。这 5 种类型分别是 GlyphRunDrawing、GeometryDrawing、ImageDrawing、VideoDrawing 和 DrawingGroup。

现在使用 WPF 的多个绘图对象来构建一幅图像,XAML 代码如下。

```xml
<Image>
    <Image.Source>
        <DrawingImage>
            <DrawingImage.Drawing>
                <!-- A group of various geometries -->
                <DrawingGroup>
                    <GeometryDrawing>
                        <GeometryDrawing.Geometry>
                            <GeometryGroup>
                                <RectangleGeometry Rect = "0,0,20,20" />
                                <RectangleGeometry Rect = "160,120,20,20" />
                                <EllipseGeometry Center = "75,75" RadiusX = "50" RadiusY = "50" />
                                <LineGeometry StartPoint = "75,75" EndPoint = "180,0" />
```

```
                </GeometryGroup>
            </GeometryDrawing.Geometry>
            <!-- A custom pen to draw the borders -->
            <GeometryDrawing.Pen>
                <Pen Thickness = "10" LineJoin = "Round" EndLineCap = "Triangle" StartLineCap =
                "Round" DashStyle = "{x:Static DashStyles.DashDotDot}" >
                    <Pen.Brush>
                        <LinearGradientBrush>
                            <GradientStop Offset = "0.0" Color = "Red" />
                            <GradientStop Offset = "1.0" Color = "Green" />
                        </LinearGradientBrush>
                    </Pen.Brush>
                </Pen>
            </GeometryDrawing.Pen>
          </GeometryDrawing>
        </DrawingGroup>
      </DrawingImage.Drawing>
    </DrawingImage>
  </Image.Source>
</Image>
</Window>
```

运行代码,页面显示效果如图 7.18 所示。从上述代码中可以看到,使用了 DrawingGroup、GeometryDrawing 和 DrawingImage 绘制图画对象。

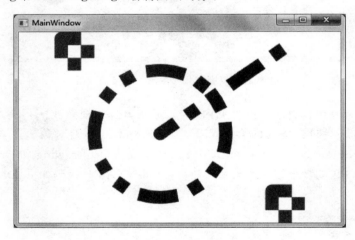

图 7.18 多类对象组合绘制图像

下面解释说明绘制图画的 5 类对象。
- GeometryDrawing:用来绘制形状。
- ImageDrawing:用来绘制图像。
- GlyphRunDrawing:用来绘制文本。
- VideoDrawing:用来播放音频或视频文件。
- DrawingGroup:绘图组,它将其他绘图组合成一个复合绘图。

Drawing 是一个通用对象。可以通过多种方式使用 Drawing 对象,上例就用 Image 和 DrawingImage 来显示为图像。

7.2 2D 图形

在 WPF 中,对 2D 图形的处理分为形状、绘制和绘制命令 3 层。其中,绘制命令是 WPF 元素合成系统的低级命令。绘制对绘制命令实施包装;形状是表现绘制的最直接的元素。

在 WPF 图形系统中有两套图形类。以 Rectangle 为例,表示 RectangleGeometry,也可以用几何路径(PathGeometry)来表示。大多数图形类位于 System.Windows.Shapes 名称空间下,与几何图形类相对应。几何图形类对应的名称空间是 System.Windows.Media。这两种图形类系统在图形系统中充当着不同的角色,功能不同。在学习中需加以区分和理解。

7.2.1 形状

形状把绘制图画带到控件中。人们熟知的 Line、Ellipse、Rectangle 和 Path 都属于几何图形的范畴。Polygon 和 Polyline 是对特别路径构建过程的简单封装。

在前面的学习中,读者已知绘制图画有其独立的坐标系统,使用偏移和变换可以在任何位置呈现图画。表 7.1 列出了形状类型及其用法。

表 7.1 形状类型及用法

形 状 类 型	用　　法
Ellipse	表示椭圆
Line	表示直线
Path	路径,可表示直线与曲线的组合结果
Polygon	一个闭合形状多边形
PolyLine	折线或未闭合的多边形
Rectangle	矩形,也可以做任意圆角矩形

在本节中,所有的元素都继承自 System.Windows.Shapes 这个名称空间。通过形状的属性给图上色。其中,Stroke 属性规定 Brush 为形状轮廓上色。表 7.2 中给出了形状属性与画笔属性的等效值及表示含义。

表 7.2 形状属性与画笔属性的等效值及表示含义

形 状 属 性	画笔属性等效值	表 示 含 义
Stroke	Brush	设置绘制形状边缘(边框)的笔刷对象
StrokeThickness	Thickness	使用设备无关单位,设置边框的宽度
StrokeLineJoin	LineJoin	确定形状拐角处的轮廓。对于没有拐角的形状,属性不起作用
StrokeMiterLimit	MiterLimit	确定形状斜角处的轮廓。对于没有拐角的形状,属性不起作用

续表

形状属性	画笔属性等效值	表示含义
StrokeDashArray	DashArray	描述形状类型轮廓的虚线和间隔的样式,常用的赋值方法可以是 StrokeDashArray="str"。str 是虚线和间隔的值的集合,奇数项为虚线长度;偶数项为间隔长度。例如,StrokeDashArray="2,1",则表示虚线长度为2,间隔为1;StrokeDashArray="2"则表示虚线和间隔都是2
StrokeDashCap	DashCap	当为Dash(虚线)时,设置的每一小段虚线两端的线帽形状
StrokeDashOffset	DashOffset	虚线开始时的偏移长度
StrokeStartLineCap	StartLineCap	决定直线开始端边缘的轮廓,只影响 Line、Polyline 及 Path
StrokeEndLineCap	EndLineCap	决定直线结束端的轮廓,只影响 Line、Polyline 及 Path

下面在 Canvas 布局中画一绿色直线(Line)、红色折线(Polyline)、不闭口的多边形(Polyline)与闭口的多边形(Polygon)。XAML 代码如下。

```
<Canvas>
    <Line Stroke="Green" X1="0" Y1="40" X2="100" Y2="40" Height="93" Width="130"
        Canvas.Left="1" Canvas.Top="1" />
    <Polyline Stroke="Red" Canvas.Left="129" Canvas.Top="12"
    Points="0,30 10,30 15,60 18,0 23,30 35,30 40,0 43,60 48,30 140,30 160,60 170,30 200,30 " />
    <Polyline Fill="Orange" Stroke="Green" StrokeThickness="2"
            Points="40,10 70,50 10,50" Canvas.Left="341" Canvas.Top="14" />
    <Polygon Fill="Orange" Stroke="Green" StrokeThickness="2"
            Points="40,10 70,50 10,50" Canvas.Left="419" Canvas.Top="14" />
</Canvas>
</Window>
```

运行上述代码,页面显示效果如图 7.19 所示。

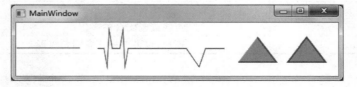

图 7.19 Line、Polyline 和 Polygon

下面在 Canvas 布局中创建圆角矩形、五角星和旋转矩形。XAML 代码如下。

```
<Canvas Height="292">
        <Rectangle Width="100" Height="50" Canvas.Left="35" Canvas.Top="33"
                Fill="Violet" RadiusX="30" RadiusY="50"/>
        <Polygon Fill="Orange" Stroke="Green" StrokeThickness="2"
            Points="10,20 100,50 10,80 66,3 66,98" Canvas.Left="204"
            Canvas.Top="12" Height="103" Width="105" />
        <Rectangle Canvas.Left="411" Canvas.Top="63" Width="40" Height="10" Fill=
"LightGreen" />
        <Rectangle Canvas.Left="411" Canvas.Top="63" Width="40" Height="10" Fill=
```

```xml
"Violet">
    <Rectangle.RenderTransform>
        <RotateTransform Angle="45"/>
    </Rectangle.RenderTransform>
</Rectangle>
<Rectangle Canvas.Left="411" Canvas.Top="63" Width="40" Height="10" Fill="Pink">
    <Rectangle.RenderTransform>
        <RotateTransform Angle="90"/>
    </Rectangle.RenderTransform>
</Rectangle>
<Rectangle Canvas.Left="411" Canvas.Top="63" Width="40" Height="10" Fill="Orange">
    <Rectangle.RenderTransform>
        <RotateTransform Angle="135"/>
    </Rectangle.RenderTransform>
</Rectangle>
<Rectangle Canvas.Left="411" Canvas.Top="63" Width="40" Height="10" Fill="Red">
    <Rectangle.RenderTransform>
        <RotateTransform Angle="180"/>
    </Rectangle.RenderTransform>
</Rectangle>
<Rectangle Canvas.Left="411" Canvas.Top="63" Width="40" Height="10" Fill="Yellow">
    <Rectangle.RenderTransform>
        <RotateTransform Angle="225"/>
    </Rectangle.RenderTransform>
</Rectangle>
<Rectangle Canvas.Left="411" Canvas.Top="63" Width="40" Height="10" Fill="blue">
    <Rectangle.RenderTransform>
        <RotateTransform Angle="270"/>
    </Rectangle.RenderTransform>
</Rectangle>
<Rectangle Canvas.Left="411" Canvas.Top="63" Width="40" Height="10" Fill="Green">
    <Rectangle.RenderTransform>
        <RotateTransform Angle="325"/>
    </Rectangle.RenderTransform>
</Rectangle>
    </Canvas>
</Window>
```

运行上述代码,页面显示效果如图 7.20 所示。代码中的第一个 Rectangle 通过 RadiusX、RadiusY 设置 X 轴与 Y 轴的半径,将矩形变成圆角。Polygon 通过定义 5 个点坐标值,产生五角星。其余的 8 个矩形通过 RotateTransform Angle 属性设置矩形的转角,生成了 8 个不同转角矩形,组成一个旋转矩形图案。

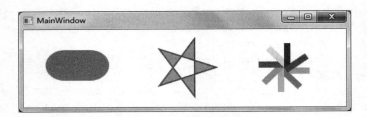

图 7.20　圆角矩形、五角星和旋转矩形

7.2.2　图像

　　图像是用于呈现矢量图和位图的数据。在前面讨论过的几何图形、笔刷、画笔、形状，这些都是用来定义矢量数据的。

　　在此，先来区分矢量图与位图。矢量图是根据几何特性来绘制图形，是用线段和曲线描述图像，矢量可以是一个点或一条线。矢量图占用空间较小，因为这种类型的图像文件包含独立的分离图像，可以自由无限制地重新组合。而位图(bitmap)，又称为光栅图和点阵图，是由许多小方格一样的像素组成的图形。两者最大的区别：矢量图形与分辨率无关，可以将它缩放到任意大小和以任意分辨率在输出设备上打印出来，都不会影响清晰度；而位图是由一个一个像素点产生的，当放大图像时，像素点也放大了，但每个像素点表示的颜色是单一的，所以在位图放大后就会出现人们平时所见到的马赛克形状。位图的文件类型为 *.bmp、*.pcx、*.gif、*.jpg、*.tif 和 *.psd(photoshop)等；矢量图形格式为 *.AI(AdobeIllustrator)、*.EPS 和 SVG、*.dwg 和 dxf(AutoCAD)和 *.cdr(Corel DRAW 的)等。

　　WPF 图像系统均针对位图格式。但是，在 DrawingImage 类型支持使用任何图画作为应用程序的图像，可看做初级的矢量图模型。下面讨论图像世界的位图图像。

1．Image 类

　　接下来，编写一个示例，在其中运用 XAML、CS 代码两种方式引用图像资源。XAML 代码如下。

```
<Grid>
    <Grid.ColumnDefinitions>
        <ColumnDefinition/>
        <ColumnDefinition/>
    </Grid.ColumnDefinitions>
    <Image Source="Images\spring.jpg" Stretch="None"/>
    <Image Name="img" Grid.Column="1" Stretch="Fill" Width="205" Height="142"/>
</Grid>
```

　　其中，<Image Source="Images\spring.jpg" Stretch="None"/>就是 XAML 代码中引用图像资源。Image 类是宿主 ImageSource 对象的一个形状。

　　在后台引用资源的 CS 代码如下。

```
public MainWindow()
{
```

```
InitializeComponent();
img.Source = new BitmapImage(new Uri("/Images/blueSky.jpg", UriKind.Relative));
}
```

运行完整的代码,页面显示效果如图 7.21 所示。

图 7.21 引用图像资源的两种方式

在后台 CS 代码中,引用图像资源的方式语句如下。

```
img.Source = new BitmapImage(new Uri("/Images/blueSky.jpg", UriKind.Relative));
```

Image 类,在此则是 ImageSource 对象的查看器。new BitmapImage 用来创建上载图片。new Uri 指路径,格式 new Uri(图片路径,路径类型),其中,路径类型可分为 UriKind.Relative、UriKind.Absolute 和 UriKind.RelativeOrAbsolute 这 3 种,它们分别代表相对路径、绝对路径和不确定路径。本节用的是相对路径。

在 Image 控件上显示一张图像时,通过设置 Stretch 属性值来表示图像的显示方式。Stretch 值可以是 None、Fill、Uniform 和 UniformToFill 这 4 种类型。如图 7.22～图 7.25 分别显示了这 4 种类型所对应的显示效果,下面来介绍这 4 种类型。

图 7.22 None

(1) None:图像以原始尺寸显示。
(2) Fill:图像缩放来适应整个空间。
(3) Uniform:图像适应整个方向的尺寸,并同时保持原始的长度比例。
(4) UniformToFill:图像适应整个控件,并同时保持原始的长度比例。

图 7.23 Fill   图 7.24 Uniform   图 7.25 UniformToFill

2. ImageSource 管道

WPF 所有的位图图像都被设计成具有一个或多个帧。例如,TIFF 和 GIF 这两图像格式文件,在单个文件中就支持多个帧。但是 PNG 和 BMP 这两种图像格式文件,则是一个帧。当加载图像时,可以使用 BitmapFrame 的静态函数 Create,下面这条 CS 语句,就是将此处所用的图片放在 F 盘的根目录。

```
BitmapFrame frame = BitmapFrame.Create(new Uri(@"F:\spring.jpg"));
```

下面使用 CroppedBitmap 类，对图片进行修剪，只显示图片的一部分。并通过 Source 属性和帧建立联系，代码如下。

```
CroppedBitmap crop = new CroppedBitmap();
crop.BeginInit();
crop.Source = frame;
crop.SourceRect = new Int32Rect(100,150,400,250);
crop.EndInit();
```

将图像转换成黑白图像，使用 FormatConvertedBitmap 类，并设置它的 Source 属性，代码如下。

```
FormatConvertedBitmap color = new FormatConvertedBitmap();
color.BeginInit();
color.Source = crop;
color.DestinationFormat = PixelFormats.BlackWhite;
color.EndInit();
```

最后要输出图像。因为在管道中的每个对象（例如，这里用到的 BitmapFrame、CroppedBitmap 和 FormatConvertedBitmap）都继承自 ImageSource，所以可在 Image 控件中使用它们中的任意一个，代码如下。输出图像如图 7.26 所示。

```
Image img = new Image();
img.Source = color;
Window w = new Window();
w.Content = img;
w.Title = "ImageSourceChannel";
w.Show();
```

若将"color.DestinationFormat = PixelFormats.BlackWhite;"中的 BlackWhite 换成 Rgb48，则图像的输出结果如图 7.27 所示。

图 7.26 通过图像管道输出黑白图片

图 7.27 通过图像管道输出 Rgb48 图片

3. Image Metadata

当前的大部分图像都支持元数据。元数据是关于数据的组织、数据域及其关系的信息。简单地说，元数据就是关于数据的数据。这里的 Image Metadata 是指图像的元数据。当用数码相机拍摄的照片，则包含了所有类型的信息。PhotoME 是一款多功能的图像元数据分析与查看工具，它能够分析和编辑所有图像的元数据，支持将数据导出，如图 7.28 所示。

图 7.28　PhotoME 对 zootopia.jpg 图的属性查看

每个 ImageSource 对象都具有 Metadata 属性,可以访问其中的信息。对于所有的位图图像 Metadata 对象都返回一个 BitmapMetadata 对象。元数据具有两个视图:简化视图和查询 API。其中,简化视图表示 BitmapMetadata 上的属性,如 CameraModel;查询 API 用于访问数据存储器的任何信息。

4. 创建图像

现实世界中的大部分图像都是使用相机或程序创建出来的,但有时,人们需要动态创建图形。动态创建图形来优化性能。例如,使用矢量和效果一次性生成复杂的图形,然后显示为位图。

生成位图图像可使用 RenderTargetBitmap 或 WritableBitmap 这两种方式。其中,RenderTargetBitmap 可以把指定的可见显示区域呈现为一个固定尺寸的位图图像,创建好位图数据,就可使用各种类型的图像编码器将其保存为图像文件,或将其显示出来。RenderTargetBitmap 继承于 ImageSource。另外,WritableBitmap 支持在位图上编辑像素点。

7.2.3　WPF 图像特效

1. 透明效果

WPF 中所有的颜色都具有一个 Alpha 值,并且每一个可视化元素都具有 Opacity(不透明)和 OpacityMask(不透明遮罩)属性。通过设置 Opacity 属性和颜色的 Alpha,可以创建出透明效果。

对于标准的 RGB 十六进制表示方法,起始的两个字符表示的是 Alpha 值。下面的示例是通过使用 LinearGradientBrush 中 Color 属性的 Alpha 值来创建透明效果,XAML 代码如下。

```
<Grid>
    <Image Source = "Images\BlackWhite.jpg"/>
    <Rectangle Width = "200" Height = "100">
        <Rectangle.Fill>
            <LinearGradientBrush EndPoint = "0,1">
                <LinearGradientBrush.GradientStops>
                    <GradientStop Offset = "0" Color = "#FFFF0000"/>
                    <GradientStop Offset = ".33" Color = "#00FFFFFF"/>
                    <GradientStop Offset = ".66" Color = "#9900FF00"/>
                    <GradientStop Offset = "0.9" Color = "#FF0000FF"/>
                </LinearGradientBrush.GradientStops>
            </LinearGradientBrush>
        </Rectangle.Fill>
    </Rectangle>
</Grid>
```

运行上述代码,页面显示效果如图 7.29 所示。其中,Grid 下面的 Image 的 Source 加载了一张黑白相间的图片,图片的作用是衬托出线性渐变笔刷的透明效果。

2. 不透明遮罩

OpacityMask(不透明遮罩)可以把透明度应用到可视化树上。需要使用径向渐变笔刷(RadialGradientBrush)对象作为不透明遮罩,XAML 代码如下。

```
<Canvas Width = "252" Height = "137" Background = "Yellow">
    <Canvas.OpacityMask>
        <RadialGradientBrush GradientOrigin = ".5 .3">
            <RadialGradientBrush.GradientStops>
                <GradientStop Offset = "0" Color = "#FF000000"/>
                <GradientStop Offset = ".33" Color = "#00000000"/>
                <GradientStop Offset = ".66" Color = "#FF000000"/>
                <GradientStop Offset = "1" Color = "#33000000"/>
            </RadialGradientBrush.GradientStops>
        </RadialGradientBrush>
    </Canvas.OpacityMask>
</Canvas>
```

运行上述代码,页面显示效果如图 7.30 所示。通过使用径向渐变笔刷对象作为 Canvas 布局中的 OpacityMask 值,产生了径向渐变笔刷不透明遮罩效果。

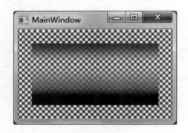
图 7.29 线性渐变笔刷设置透明效果

图 7.30 不透明遮罩效果

3. BitmapEffects

OpacityMask 是通过合成引擎对像素的生成过程进行编辑,而 BitmapEffects 是针对每个像素进行操作,继承自 UIElement,被称为位图效果(Bitmap Effects)。BitmapEffects 是源于它在合成引擎所生成的位图(真实的像素)上进行操作。虽然所有的效果都能被应用于任何元素上,但其中一些效果(DropShadowBitmapEffect)更适合矢量内容。

在 WPF 中,使用 BitmapEffect 对所有 Visual 对象进行位图特效处理。它是基于像素级别的,而且是基于软件处理模式而非硬件加速的处理模式。常见的位图特效处理有斜面特效(BevelBitmapEffect)、虚化效果(BlurBitmapEffect)、外辉光效果(OuterGlowBitmapEffect)、阴影效果(DropShadowBitmapEffect)和浮雕特效(EmbossBitmapEffect)。因为软件处理模式的效率相对较低,位图效果是非常消耗 CPU 资源的。因为它是由 CPU 计算的而不是 GPU(Graphics Processing Unit,图形处理器),而且不要将位图效果与动画(Animation)一起使用,它常常使动画变得很不流畅。故此,在有大量 Visual 对象或存在有大量动画时慎用。

7.3 3D 图形

在 WPF 中,3D 图形是用矢量图形表示的一种形式,并不具备物理模型、碰撞检测及用于编写游戏需要的 3D 环境所需的高级服务。

WPF 生成三维图形的基本思想是能得到一个物体的三维立体模型(Model)。由于计算机的屏幕是二维的,因此需要定义了一个用于给物体拍照的照相机(Camera)。拍到的照片其实是物体到一个平坦表面的投影。这个投影由 3D 渲染引擎渲染成位图。引擎通过计算所有光源对 3D 空间中物体的投影面反射的光量,来决定位图中每个像素点的颜色。

在 WPF 3D 中,需要理解的基本概念有三维空间坐标系、模型(Models)、材质(Materials)、光源(Lights)、照相机(Cameras)和变换(Transform)。

7.3.1 WPF 坐标系

当使用 WPF 创建图形时,应该清楚地知道图形显示在什么地方,要明白这一点,就需要对 WPF 中的坐标系统有一定的认识。

在 WPF 二维坐标系中,左上角是坐标原点,向右为 X 轴的正方向,向下为 Y 轴的正方向,如图 7.31 所示的 WPF 默认二维坐标系。

在 WPF 的三维坐标系中,原点一般位于在 WPF 中创建的三维对象的中心。三维坐标的 X 轴正方向朝右,Y 轴的正方向朝上,Z 轴的正方向从原点向外朝向观察者。图 7.32 给出了 WPF 三维空间坐标系。

充分理解 WPF 的坐标系,在 Canvas 布局控件中画一条直线,采用系统默认的二维坐标系,XAML 代码如下。

```
< Canvas Height = "500" Width = "500">
    < Line X1 = "0" Y1 = "0" X2 = "200" Y2 = "100" Stroke = "Black" StrokeThickness = "2" />
</Canvas>
```

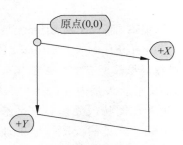
图 7.31　WPF 默认二维坐标系

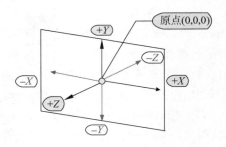
图 7.32　WPF 三维坐标系统

运行上述代码,页面显示效果如图 7.33 所示。WPF 除了默认的坐标系统外,也支持用户自定义坐标系统。例如,在很多场景中,是以左下角为坐标原点,Y 轴方向指向上方。图 7.34 所示为用户自定义二维坐标系。

图 7.33　WPF 默认二维坐标系

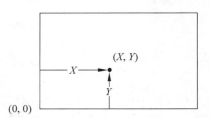
图 7.34　用户自定义二维坐标系

在用户自定义坐标系中,画一条从原点出发的直线,XAML 代码如下。

```xml
<!-- 保留 Window 代码部分 -->
Title = "MainWindow" Height = "350" Width = "525">
    <Border BorderBrush = "Black" BorderThickness = "1" Height = "130" Width = "200">
        <Canvas Height = "272" Width = "244">
            <Canvas.RenderTransform>
                <TransformGroup>
                    <ScaleTransform ScaleY = " - 1" />
                    <TranslateTransform Y = "200" />
                </TransformGroup>
            </Canvas.RenderTransform>
            <Line X1 = "0" Y1 = "0" X2 = "150" Y2 = "100" Stroke = "Black" StrokeThickness = "2"
                Canvas.Left = "1" Canvas.Top = "71" Height = "110" Width = "175" />
        </Canvas>
    </Border>
</Window>
```

运行上述代码,页面显示效果如图 7.35 所示。这时,在变换了的二维坐标系下,按正常方式,在 Canvas 中添加一个按钮和文本框,XAML 代码如下。

```xml
<Button Canvas.Top = "144" Canvas.Left = " - 1" Foreground = "Green" Content = "Button" />
<TextBlock Canvas.Top = "179" Canvas.Left = "1" Foreground = "Blue" Text = "TextBlock" />
```

运行上述代码,页面显示效果如图 7.36 所示。发现文本和按钮是颠倒的。要解决这个问题,在按钮和文本框上运用变换的 XAML 如下。

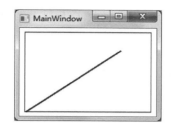
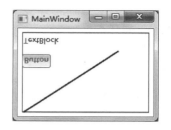

图7.35　变换二维坐标系下的直线　　　图7.36　变换二维坐标系下的按钮和文本框

```
< Button Canvas.Top = "144" Canvas.Left = " - 1" Foreground = "Green" Content = "Button" >
    < Button.RenderTransform >
        < ScaleTransform ScaleY = " - 1" />
    </Button.RenderTransform >
</Button >
< TextBlock Canvas.Top = "179" Canvas.Left = "1" Foreground = "Blue" Text = "TextBlock">
    < TextBlock.RenderTransform >
        < ScaleTransform ScaleY = " - 1" />
    </TextBlock.RenderTransform >
</TextBlock >
```

运行上述代码,页面显示效果如图7.37所示。按钮与文本框控件正常显示。

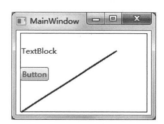

图7.37　按钮和文本框变回正常输出效果

7.3.2　模型

在3D图形的世界中,所有物体都被描述成为一系列三角形的集合。三角形是用来描述平面最小的几何体。在场景中,渲染引擎可以计算出每一个三角形的颜色,这取决于它的材质和它与光线的角度。用三角形构建3D世界,这些点都在同一平面上,这种表面计算和渲染起来相对简单。

WPF的所有显示都是用3D管道来实现的。所有的控件、形状、文本、绘制图画都能被呈现为3D三角形。一个3D模型定义了场景中的一个物体,它包含一个Geometry对象,这个对象是一个网格(Mesh)和一个材质(Material)。也可以让物体的每一个表面都具有一种材质(Material)和一个笔刷(Brush)。材质定义了一个具体角度的光的反射量,而笔刷定义了表面的颜色。笔刷可以是一种单纯的颜色,也可以是渐变的,甚至可以是一幅图片,这些称为纹理(Texture)。

3D物体的一个表面称为一个网格(Mesh)。一个网格被定义为许多3D点,这些点称为顶点（Vertices）。这些顶点通过环绕方式连接起来形成三角形。每一个三角形都有一个正面和反面，只有正面才会被渲染。三角形的正面可以通过点的环绕顺序来确定。

在此，先来绘制一个简单的三角形。其 XAML 代码如下。

```xml
<!-- 保留 Window 代码部分 -->
Title = "MainWindow" Height = "350" Width = "525">
    <Grid>
        <Viewport3D>
            <Viewport3D.Camera>
                <PerspectiveCamera Position = " - 2,2,2" LookDirection = "2, - 2, - 2"
                    UpDirection = "0,1,0"/>
            </Viewport3D.Camera>
            <ModelVisual3D>
                <ModelVisual3D.Content>
                    <GeometryModel3D>
                        <GeometryModel3D.Geometry>
                            <MeshGeometry3D Positions = " 1,0,0 0, - 1,0 0,0,1 "/>
                        </GeometryModel3D.Geometry>
                        <GeometryModel3D.Material>
                            <DiffuseMaterial Brush = "Green" />
                        </GeometryModel3D.Material>
                    </GeometryModel3D>
                </ModelVisual3D.Content>
            </ModelVisual3D>
        </Viewport3D>
    </Grid>
</Window>
```

运行上述代码，页面显示效果如图 7.38 所示。其中 Viewport3D 是连接 2D 和 3D 世界大门的一个控件。

每一个 3D 场景都有一个摄像机。摄像机定义 Position、LookDirection 和 UpDirection 属性。WPF 支持正交（Orthographical）和透视（Perspective）摄像机。其中 PerspectiveCamera（透视相机）指定 3D 模型到 2D 可视图面的投影。换言之，它描述各个面均聚集到某个水平点的平截体。对象离摄像机越近就显得越大，离得越远则显得越小。

在本例中，尽管 < DiffuseMaterial Brush = " Green " /> 通过笔刷定义 Material 为绿色，但是运行结果是黑色。因为模型中没有加入光照。在 < ModelVisual3D.Content > 后加入光照效果的 XAML 代码，再运行程序，页面显示效果如图 7.39 所示。三角形变成期望的绿色。

```xml
< ModelVisual3D.Content >
< Model3DGroup >
    < AmbientLight  Color = "White" />
    <!-- 保留与上面相同的代码部分 -->
</Model3DGroup >
</ModelVisual3D.Content >
```

图 7.38　WPF 3D 三角形　　　　图 7.39　加入光照后的三角形的 3D 模型

从上面的示例中可知，GeometryModel3D.Geometry 和 MeshGeometry3D 总是连在一起来创建模型。其中，使用 MeshGeometry3D 定义模型时，首先在 WPF 三维坐标系中定义一系列点的位置，然后从这些方位列表中选择 3 个点来定义三角形，再来设置纹理坐标，把这些点映射到材质上。为了理解它们之间的关系，来创建一个三面立方体。XAML 代码如下。

```xml
<!-- 保留 Window 代码部分 -->
<Grid>
    <Viewport3D Width="200" Height="200">
        <Viewport3D.Camera>
            <PerspectiveCamera LookDirection="-.7,-.8,-1"
                               Position="3.8,4,4"
                               FieldOfView="17"
                               UpDirection="0,1,0"/>
        </Viewport3D.Camera>
        <ModelVisual3D>
            <ModelVisual3D.Content>
                <Model3DGroup>
                    <PointLight Position="3.8,4,4"
                                Color="White"
                                Range="8"
                                ConstantAttenuation="1.0"/>
                    <GeometryModel3D>
                        <GeometryModel3D.Geometry>
                            <MeshGeometry3D
                    TextureCoordinates="0,0 1,0 0,1 1,1 0,0 1,1 1,1,1"
                    Positions="0,0,0 1,0,0 0,1,0 1,1,0 0,1,-1 1,1,-1 1,0,-1"
                    TriangleIndices="0,1,2 3,2,1 4,2,3 5,4,3 5,3,1 6,5,1 "/>
                        </GeometryModel3D.Geometry>
                        <GeometryModel3D.Material>
                            <DiffuseMaterial>
                                <DiffuseMaterial.Brush>
                                    <LinearGradientBrush EndPoint="1,0">
                                        <LinearGradientBrush.GradientStops>
                                            <GradientStop Offset="0" Color="Black" />
                                            <GradientStop Offset="1" Color="White"/>
                                        </LinearGradientBrush.GradientStops>
                                    </LinearGradientBrush>
```

```
            </DiffuseMaterial.Brush>
          </DiffuseMaterial>
        </GeometryModel3D.Material>
      </GeometryModel3D>
    </Model3DGroup>
   </ModelVisual3D.Content>
  </ModelVisual3D>
 </Viewport3D>
</Grid>
</Window>
```

运行上述代码,页面显示效果如图 7.40 所示。分析上述创建立方体的 XAML 代码,涉及 3D 立方体的创建步骤和 MeshGeometry3D 的常用属性。下面分别介绍它们。

创建 3D 立方体的步骤如下:

(1) 首先用 Positions 属性定义一系列点。立方体具有 8 个顶点。由于此处立方体只有 3 个可见的面,有一个点是不可见点。所以点的序号从 0 开始,从第 0 点到第 6 点,共需要 7 个点。图 7.41(a)标出 7 个点分别在三维坐标系中的位置。

图 7.40 带有线性热渐变色三面立方体

其中,上述代码中的"Positions="0,0,0 1,0,0 0,1,0 1,1,0 0,1,−1 1,1,−1 1,0,−1""就是指从第 0 点到第 6 点所对应的三维坐标值。

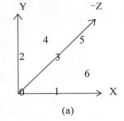 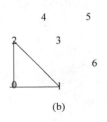 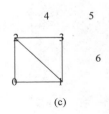

(a)　　　　　　　　　(b)　　　　　　　　(c)

图 7.41　3D 立方体构建步骤图

(2) 设定网状结构,有序串联这些点,构建三角形。构建第一个三角形,如图 7.41(b)所示,它是由 0、1、2 这 3 个点组成的。继续构建第二个三角形,如图 7.41(c)所示,它是由 1、2、3 这 3 个点组成的。以此类推,第三个、第四个、第五个、第六个三角形的构建过程如图 7.42 所示。

图 7.42　3D 立方体内三角形构建顺序图

(3) 将这些点映射到材质上(或称为纹理"Texture")。

接下来介绍 MeshGeometry3D 常用的 4 个属性,如表 7.3 所示。

表 7.3　MeshGeometry3D 常用属性及其描述

属　　性	描　　述
Normals	获取或设置 MeshGeometry3D 的法向量的集合
Positions	获取或设置 MeshGeometry3D 的顶点位置的集合
TextureCoordinates	获取或设置 MeshGeometry3D 的纹理坐标集合
TriangleIndices	获取或设置 MeshGeometry3D 的三角形索引的集合

初学者不易区分 TriangleIndices 和 Positions 的关系。因为在 3D 图形的世界中,所有物体都可以被描述成为一系列三角形的集合。以上面的"TriangleIndices＝"0,1,2 3,2,1 4,2,3 5,4,3 5,3,1 6,5,1""为例,它表示三角形索引的集合。其中,第一个三角形是由 0、1 和 2 这 3 个点构成;第二个三角形是由 3、2 和 1 这 3 个点构成;以此类推。这里面的每个数字对应图 7.41 中的每个点。这种对应关系呈现三角形的不同面。可以看出,上面每三个点组成的一个三角形都是逆时针顺序的,这是因为 WPF 采用逆时针的环绕方式来显示正面,或者用右手定则:握住右手,伸出拇指,其余四指为逆时针方向,此时拇指指向的是 WPF 正面。

TextureCoordinates:纹理坐标,用于确定将 Material 映射到构成网格的三角形的顶点的方式。(0,0)代表整个图形的左上角;(1,1)则代表右下角;(0,1)则代表左下角;(1,0)则代表右上角。

Normals:法向量,是与定义网格的每个三角形的面垂直的向量。法向量用于确定是否点亮给定三角形面。如果指定三角形索引,则将考虑相邻面来生成法向量。

7.3.3　材质

一个 3D 模型定义了场景中的一个物体。它包含一个 Geometry 对象,这个对象可以看作由一个网格(Mesh)和一个材质(Material)构成。材质本身具有一个笔刷。三维对象使用的材料分为漫射材料(Diffuse)、放射材料(Emmisive)和反射材料(Specular)三类,它们的特性如下。

(1) 漫射材料:确定三维对象在直射光(白光)照射下的颜色,其作用就如同墙面喷漆一样。

(2) 放射材料:使对象产生发光效果。光的颜色由材料的颜色决定。

(3) 反射材料:控制三维对象上高光反射区域的颜色。高光反射区域指在金属铬等光滑亮泽表面上看到的光亮区域。

对上述立方体选用 ImageBrush 作为材质,XAML 代码如下。

```
<GeometryModel3D.Material>
    <DiffuseMaterial>
        <DiffuseMaterial.Brush>
            <ImageBrush ImageSource = "Images\spring.jpg"/>
        </DiffuseMaterial.Brush>
    </DiffuseMaterial>
</GeometryModel3D.Material>
```

运行上述代码,页面显示效果如图 7.43 所示。ImageBrush 使用图片做立方体材质。

7.3.4 光源与照相机

1. 光源

没有光源什么也看不到。因此对于 3D 图形，首先需要在场景中至少放置一个光源来照亮模型。WPF 中有 4 种光源类型分别是 AmbientLight（环境光）、SpotLight（聚焦光）、DirectionalLight（方向光）和 PointLight（点光）。

将图 7.43 的效果图更换成环境光做光源，XAML 代码为< AmbientLight Color="White" />，页面显示效果如图 7.44 所示。图 7.45 给出了不同类型光源的投影效果。

图 7.43　图片做立方体材质　　　　图 7.44　环境光照下的立方体

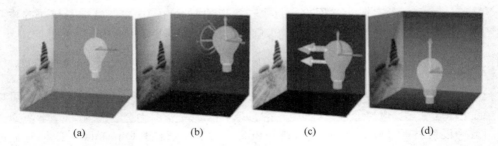

图 7.45　不同类型光源的投影效果

这 4 种光源发光效果不同，各自特征如下。

（1）环境光：将光投向各个方向，使所有对象均匀受光，如图 7.45(a)所示。如果只用环境光，则对象可能会显得褪色，而且颜色单一。为了获得最佳效果，需要使用其他光。

（2）聚焦光：投射光所投射的光如同聚光灯一般，光从发光位置发出，并在锥形区域内传播。投射光不影响到位于锥形发光区域以外的那部分三维对象，如图 7.45(b)所示。

（3）定向光：沿着特定的方向均匀平行投射，就像太阳光一样，如图 7.45(c)所示。

（4）点光：从一个点向所有方向投射光，就像普通的灯泡一样，如图 7.45(d)所示。

2. 照相机

照相机是观察者观察三维对象形态和位置的工具，照相机的位置坐标及与对象的距离直接影响到三维对象的呈现。WPF 中的相机有正交相机（OrthographicCamera）和透视（PerspectiveCamera）相机两类。其中，OrthographicCamera（正交相机）是指定的三维模型到二维可视化图面上的正投影。当然，正交相机也需要指定位置，如观测时的"向上"方向等属性。它描述的并不包括透视的收缩投影（也就是说，如果从观察者的角度来说，前者对观察对象没有透视感,因为描述的是和侧面平行的取景框。透视相机的工作原理与普通照相机镜头类似,对象离照相机越远,看起来就越小,观察到的对象则有远小近大的效果。

下面以 PerspectiveCamera(透视相机)为例,查阅相机的属性值的含义。常用的属性及其描述如表 7.4 所示。

表 7.4　PerspectiveCamera 常用属性及其描述

属　　性	描　　述
Position	获取或设置以世界坐标表示的相机位置
FieldOfView	获取或设置一个值,该值表示相机的水平视角
LookDirection	获取或设置定义相机在世界坐标中的拍摄方向的 Vector3D
UpDirection	获取或设置定义相机向上方向的 Vector3D
NearPlaneDistance	获取或设置一个值,该值指定到相机近端剪裁平面的摄像机的距离
FarPlaneDistance	获取或设置一个值,该值指定到相机远端剪裁平面的摄像机的距离

当然,照相机的位置坐标是可以变化的,还以透视相机为例,调整相机位置,如图 7.46 所示。当相机位置靠近 Z 轴中心时,即 Z 坐标值变小,观察到的对象变大;当相机位置远离 Z 轴中心时,即 Z 坐标值变大,观察到的对象变小。相机的位置就是观察者的位置,相机的位置可以任意设置,这样观察到的三维对象的"形象"就会发生变化。在三维场景中正确设置相机位置很重要。

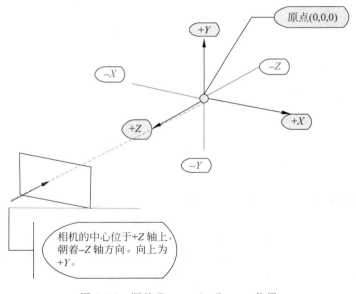

图 7.46　调整 PerspectiveCamera 位置

7.3.5　变换

变换(Transform)是使用坐标系统来改变形状或元素的绘制方式。在 WPF 中,变换是由继承自 System.Windows.Media.Transform 抽象类的类表示,表 7.5 列出常用的变换类及其重要属性。

表7.5 常用的变换类及其重要属性

名称	描述	重要属性
TranslateTransform	将坐标系移动一定的距离	X、Y
RotateTransform	旋转坐标系，绕着选择的中心旋转	Angle、CenterX、CenterY
ScaleTransform	放大或缩小坐标系，实现图形的缩放	ScaleX、ScaleY、CenterX、CenterY
SkewTransform	让坐标系倾斜。例如，正方形变成平行四边形	AngleX、AngleY、CenterX、CenterY
MatrixTransform	通过矩阵乘积修改坐标系	Matrix
TransformGroup	组合多个变换，使用应注意变换顺序	N/A

由表7.3可知，所有变换都使用数学中的矩阵改变形状坐标。让Canvas中的Rectangle旋转25°，XAML代码如下。

```
<Canvas>
    <Rectangle Width = "70" Height = "12" Stroke = "Green" Fill = "Blue"
        Canvas.Left = "100" Canvas.Top = "100">
        <Rectangle.RenderTransform>
            <RotateTransform Angle = "25" CenterX = "45" CenterY = "5"/>
        </Rectangle.RenderTransform>
    </Rectangle>
</Canvas>
```

7.4 小结

WPF基于元素合成的设计思想也适用于WPF的图形系统。本章从常用的几何图形元素出发，学习了绘制图画和2D形状及属性，并讲解了WPF 3D的三维空间坐标系、模型（Models）、材质（Materials）、光源（Lights）、照相机（Cameras）和变换（Transform）。

习题与实验7

1. 根据7.1节的内容，熟练掌握LineSegment、BezierSegment、ArcSegment、CombineGeometry和GeometryGroup这些混合几何图形，按如下要求绘图。

(1) 用CombineGeometry绘制两个等大小的焦点分别在X轴、Y轴的椭圆，填充规则为Xor，页面效果如图7.47所示。

(2) 用GeometryGroup绘制5个大小相同的圆，填充规则为EvenOdd，页面显示效果如图7.48所示。

2. 在Canvas布局中，采用Polyline绘制心电图。

3. 使用MeshGeometry3D定义模型，创建一个三棱锥，选用ImageBrush作为材质，实现如图7.49所示的环境光照下的三棱锥。

图7.47 Xor填充双椭圆

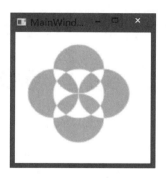 

图 7.48　EvenOdd 填充 5 个同心圆　　　图 7.49　环境光照下的三棱锥

第 8 章

动画与媒体

媒体是传播信息的媒介。生活中常见的媒体类型是音频和视频。动画是视频媒体的一部分,它依靠人眼的视觉残留效应,将静止的画面变为动态的图像技术。

8.1 动画基础

在以往的程序开发中,如果想构建动画,需要定时器和自定义的绘图元素,并让这些绘图元素根据定时器做出相应的改变,以实现动画效果,开发难度和工作量都是很高的。并且这些动画的拓展性和灵活性一般很弱,代码量和复杂度却很大。而在 WPF 中,可以使用声明的方式构建动画,甚至不需要任何后台代码,就可以实现动画效果。WPF 提供的动画模型和强大的类库,让一般动画的实现都变得轻而易举。在 WPF 中,创建更加复杂的动画,甚至也可以使用设计工具或第三方工具在 XAML 中实现。

8.1.1 动画的概念

世界著名动画艺术家、英国人 John Hales 说:"运动是动画的本质。"也有人说"动画是运动的艺术"。总之,动画与运动是分不开的。

动画是集绘画、漫画、电影、数字媒体、摄影、音乐、文学等众多艺术门类于一身的艺术表现形式。最早发源于 19 世纪上半叶的英国,兴盛于美国,中国动画起源于 20 世纪 20 年代。动画是一门年青的艺术,它是唯一有确定诞生日期的一门艺术。1892 年 10 月 28 日埃米尔·雷诺首次在巴黎著名的葛莱凡蜡像馆向观众放映光学影戏,标志着动画的正式诞生,同时埃米尔·雷诺也被誉为"动画之父"。

动画的英文有很多表述,如 animation、cartoon、animated cartoon。其中较正式的"Animation"一词源自于拉丁文字根 anima,意思为"灵魂",动词 animate 是"赋予生命"的意思,引申为使某物活起来的意思。所以动画可以定义为使用绘画的手法,创造生命运动的艺术。

8.1.2 动画的原理

当人们在电影院看电影或在家里看电视时,画面中的人物动作是流畅的、自然的和连续的。但是当仔细看一段电影胶片时,看到的画面却不连续。只有以一定的速率投影在银幕上才有运动的视觉效果,这种现象可以由视觉暂留的原理来解释。

视觉暂留现象即视觉暂停现象(Persistence of vision,Visual staying phenomenon,duration of vision)又称"余晖效应",1824 年由英国伦敦大学教授皮特·马克·罗葛特在他

的研究报告《移动物体的视觉暂留现象》中最先提出。

物体在快速运动时,当人眼所看到的影像消失后,人眼仍能继续保留其影像 0.1~0.4 秒的图像,这种现象被称为视觉暂留现象,是人眼具有的一种性质。人眼观看物体时,成像于视网膜上,并由视神经输入大脑,感觉到物体的像。但当物体移去时,视神经对物体的印象不会立即消失,而要延续 0.1~0.4 秒的时间,人眼的这种性质被称为"眼睛的视觉暂留"。

人眼在观察景物时,光信号传入大脑神经,需经过一段短暂的时间,光的作用结束后,视觉形象并不立即消失,这种残留的视觉称"后像",视觉的这一现象则被称为"视觉暂留"。

视觉暂留现象是光对视网膜所产生的视觉在光停止作用后,仍保留一段时间的现象,其具体应用是电影的拍摄和放映。原因是由视神经的反应速度造成的,其值是 1/24 秒,这是动画和电影等视觉媒体形成和传播的根据。视觉实际上是靠眼睛的晶状体成像,感光细胞感光,并且将光信号转换为神经电流,传回大脑引起人体视觉。感光细胞的感光是靠一些感光色素,感光色素的形成是需要一定时间的,这就形成了视觉暂停的机制。

视觉暂留现象首先被中国人运用,宋朝时走马灯便是据历史记载中最早的视觉暂留运用。随后法国人保罗·罗盖在 1828 年发明了留影盘,它是一个被绳子在两面穿过的圆盘。盘的一个面画了一只鸟,另一面画了一个空笼子。当圆盘旋转时,鸟在笼子里出现了,这证明了当眼睛看到一系列图像时,它一次保留一个图像。

在动画创作时,以每秒 5~30 幅的速度得到动画场景(包含运动物体)瞬间的若干幅静止图片,每幅静止的图片被称为一帧,然后按照动作发生的时间顺序,以相同速度播放这些图片,利用人眼视觉暂留特性,重新看到运动场景。

8.1.3 传统动画与 WPF 动画

在以往的程序开发中,如果想构建动画,需要定时器和自定义的绘图元素,并让这些绘图元素根据定时器做出相应的改变,以实现动画效果,开发难度和工作量都是很高的。并且这些动画的拓展性和灵活性一般很弱,代码量和复杂度却很大。

下面实现一个按钮宽度变化的动画,在传统意义上实现该动画,则后台 CS 代码如下。

```
public partial class MainWindow : Window
{
    public MainWindow()
    {
        InitializeComponent();
    }
    private void Window_Loaded(object sender, RoutedEventArgs e)
    {
        var timer = new System.Windows.Threading.DispatcherTimer();
        timer.Tick += new EventHandler(OnTimedEvent);
        timer.Interval = TimeSpan.FromSeconds(1.0/20);
        timer.Start();
    }
    int index = 0;
    private void OnTimedEvent(object sender, EventArgs e)
    {
        index++;
```

```
        if (index > 30)
            index = 0;
        button1.Width = 8 * (index++);
    }
}
```

这段代码的功能是每隔 1/20 秒更新一次按钮(button1)的宽度,在 2s 内将其宽度从 0 变为 240,重复播放。这是通过更新计时器的传统动画制作方法。在 WPF 中,实现该动画的 CS 代码如下。

```
private void Window_Loaded(object sender, RoutedEventArgs e)
{
            var widthAnimation = new DoubleAnimation()
            {
                From = 0,
                To = 240,
                Duration = TimeSpan.FromSeconds(2),
                RepeatBehavior = RepeatBehavior.Forever,
            };
            button1.BeginAnimation(WidthProperty, widthAnimation);
}
```

传统动画与 WPF 动画相比较而言,WPF 的动画的实现方式更简洁,可以与 XAML 无缝集成、运行起来比传统动画更流畅,这是因为传统动画的精度不够高、帧率受限制。下面分析传统动画的精度与帧率。

传统动画在 UI 线程上修改 UI 控件的属性(按钮的宽度),因此 DispatcherTimer 操作与其他操作一样,需要放置到 Dispatcher 队列中,它并不能保证恰好在该时间间隔中。从某种意义上讲,它不适合动画这种间隔很短的计时。

动画的流畅性一般取决于每秒更新的帧数,也就是常说的帧率。人眼睛上限是 70 帧,而上述传统动画代码是每秒 20 帧的帧率。而 WPF 动画是根据计算机的性能和当前进程的繁忙程度尽可能地增大帧率,WPF 动画是远大于 20 帧,因此要流畅得多。

鉴于 WPF 动画的优势,读者还需要了解 WPF 的动画类型。

8.2 动画类型

WPF 动画在 System.Windows.Media.Animation 名称空间中,该名称空间包含 3 种动画类型,它们分别是线性插值动画(17 个)、关键帧动画(22 个)和路径动画(3 个)。

8.2.1 线性插值动画

线性插值动画表现为元素的某个属性在开始值和结束值之间逐步增加的线性过程。在 WPF 动画的名称空间中,有 17 个线性插值动画类。这些类名由两个单词构成,前面的单词表示线性插值的类型名,第二个单词是动画(Animation)。这些线性插值动画类分别是 ByteAnimation、ColorAnimation、DecimalAnimation、DoubleAnimation、Int16Animation、Int32Animation、Int64Animation、Point3DAnimation、PointAnimation、QuaternionAnimation、

RectAnimation、Rotation3DAnimation、SingleAnimation、SizeAnimation、ThicknessAnimation、Vector3DAnimation 和 VectorAnimation。

现在利用 DoubleAnimation，实现文字的淡入效果，XAML 代码如下。

```
<!-- 保留 Window 代码部分 -->
Title = "MainWindow" Height = "350" Width = "525" Loaded = "Window_Loaded" >
    < Grid >
        < TextBlock Height = "50" Width = "220" Foreground = "Green" FontSize = "36"
                Name = "textBlock1" Text = "文字淡入效果"/>
    </Grid>
</Window>
```

实现文字的淡入效果动画，CS 代码如下。

```
public partial class MainWindow : Window
{
    public MainWindow()
    {
        InitializeComponent();
    }
    private void Window_Loaded(object sender, RoutedEventArgs e)
    {
        var da = new DoubleAnimation()
        {
            From = 0, //起始值
            To = 1, //结束值
            Duration = TimeSpan.FromSeconds(3), //动画持续时间
            RepeatBehavior = RepeatBehavior.Forever, //动画行为状态
        };
        this.textBlock1.BeginAnimation(TextBlock.OpacityProperty, da);
        //开始动画
    }
}
```

运行上述代码，页面显示效果如图 8.1 所示。

再利用 ThicknessAnimation，实现文字平移效果，XAML 代码如下。

图 8.1　文字淡入动画效果

```
<!-- 保留 Window 代码部分 -->
Title = "MainWindow" Height = "350" Width = "525" Loaded =
"Window_Loaded">
    < Grid >
        < TextBlock Height = "50" Foreground = "Green" FontSize = "36"
          Name = "textBlock1" Text = "文字平移" Margin = "0,21,0,240" />
    </Grid>
</Window>
```

实现文字平移效果动画，CS 代码如下。

```
private void Window_Loaded(object sender, RoutedEventArgs e)
{
```

```
            var ta = new ThicknessAnimation()
            {
                From = new Thickness(0,0,100,0),              //起始值
                To = new Thickness(100, 0, 0,0),              //结束值
                Duration = TimeSpan.FromSeconds(3),           //动画持续时间
            };
            this.textBlock1.BeginAnimation(TextBlock.MarginProperty, ta); //开始动画
}
```

运行上述代码，文字平移初始值页面显示效果如图 8.2 所示，文字平移结束值页面显示效果如图 8.3 所示。

 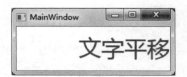

图 8.2　文字平移初始值　　　　　　　图 8.3　文字平移结束值

8.2.2　关键帧动画

关键帧动画是以时间为结点，在指定时间结点上，让属性达到某个值。WPF 动画的名称空间中共有 22 个关键帧动画类。BooleanAnimationUsingKeyFrame 是布尔关键帧动画，分析这个关键帧动画的命名可知，WPF 关键帧动画类名由三部分组成。类名的第一部分是关键帧的类型名；第二部分是 Animation；第三部分是 UsingKeyFrames。命名规则：数据类型名＋AnimationUsingKeyFrames。下面只给出 WPF 的 22 个关键帧动画类名的数据类型名，分别是 Boolean、Bytes、Char、Color、Decimal、Double、Int16、Int32、Int64、Matrix、Object、Point3D、Point、Quaternion、Rect、Rotation3D、Single、Size、String、Thickness、Vector3D 和 Vector。

现在利用 DoubleAnimationUsingKeyFrames 关键帧动画实现一个弹跳的矩形框，代码中还用到事件触发器（EventTrigger）来监视事件。事件触发器的作用是当一个事件发生时，其引发相关的动画来响应。XAML 代码如下。

```xml
<Canvas>
    <Rectangle Fill = "Gray" Width = "15" Height = "15">
        <Rectangle.Triggers>
        <!-- 当矩形加载后,事件触发器,引发动画 -->
            <EventTrigger RoutedEvent = "Rectangle.Loaded">
                <BeginStoryboard>
                    <Storyboard>
                    <DoubleAnimation From = "0" To = "800" Duration = "0:0:10"
                        Storyboard.TargetProperty = "(Canvas.Left)"
                        RepeatBehavior = "Forever" AutoReverse = "True"/>
                    <DoubleAnimationUsingKeyFrames Duration = "0:0:2"
                        Storyboard.TargetProperty = "(Canvas.Top)" RepeatBehavior = "Forever">
                        <DoubleAnimationUsingKeyFrames.KeyFrames>
                            <LinearDoubleKeyFrame Value = "0" KeyTime = "0:0:0"/>
```

```
                    < LinearDoubleKeyFrame Value = "50" KeyTime = "0:0:0.5"/>
                    < LinearDoubleKeyFrame Value = "200" KeyTime = "0:0:1"/>
                    < LinearDoubleKeyFrame Value = "50" KeyTime = "0:0:1.5"/>
                    < LinearDoubleKeyFrame Value = "0" KeyTime = "0:0:2"/>
                </DoubleAnimationUsingKeyFrames.KeyFrames >
              </DoubleAnimationUsingKeyFrames >
            </Storyboard >
          </BeginStoryboard >
        </EventTrigger >
      </Rectangle.Triggers >
    </Rectangle >
  </Canvas >
</Window >
```

这里有两个移动的时间轴。第一个移动矩形从左到右，使用常规的 DoubleAnimation；第二个通过使用 DoubleAnimationUsingKeyFrames 控制了 5 个帧垂直的位置，如图 8.4 所示，5 个关键帧显示了这个矩形跳动的顶部和底部及中途点。

代码中每个关键帧的值都使用 LinearDoubleKeyFrame 线性添加方式，改变的速度是介于两个帧之间的常量。这就引起运动轨迹不是特别平滑的现象，如图 8.5 所示。为了增强动画的流畅性，可以通过添加更多的关键帧，但是这里使用曲线插值关键帧，如图 8.6 所示，提高平滑度，而不需要添加更多的关键帧。在上述代码的基础上，后 4 个线性插值关键帧（LinearDoubleKeyFrame）变为曲线插值关键帧（SplineDoubleKeyFrame），变动的 XAML 代码如下。

```
< DoubleAnimationUsingKeyFrames.KeyFrames >
    < LinearDoubleKeyFrame Value = "0" KeyTime = "0:0:0"/>
    < SplineDoubleKeyFrame Value = "50" KeyTime = "0:0:0.5"/>
    < SplineDoubleKeyFrame Value = "200" KeyTime = "0:0:1"/>
    < SplineDoubleKeyFrame Value = "50" KeyTime = "0:0:1.5"/>
    < SplineDoubleKeyFrame Value = "0" KeyTime = "0:0:2"/>
</DoubleAnimationUsingKeyFrames.KeyFrames >
```

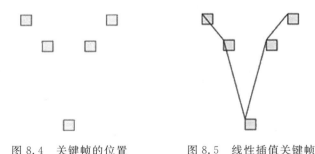
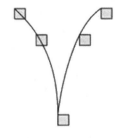

图 8.4　关键帧的位置　　　图 8.5　线性插值关键帧　　　图 8.6　曲线插值关键帧

8.2.3　路径动画

路径动画的表现方式是修改数值使其符合 PathGeometry 对象描述的形状，并且让元素沿着路径移动。虽然人们也可以通过控制动画的旋转和偏移实现对象的移动，但路径动画更专业，它的实现更加简洁明了。WPF 中有 3 个路径动画类，它们分别是

WPF 编程基础

DoubleAnimationUsingPath、MatrixAnimationUsingPath 和 PointAnimationUsingPath。

路径动画中最常用的是 MatrixAnimationUsingPath 路径动画,它用来控制对象的 MatrixTransform。

现在让一个矩形沿着一个椭圆形路径轨迹移动,在 XAML 中放置一个 Button 来控制动画启动,XAML 代码如下。

```xml
<!-- 保留 Window 代码部分 -->
<Grid Height="300" Width="489">
    <Path x:Name="path1" Stroke="Blue" Margin="0,0,0,43">
        <Path.Data>
            <EllipseGeometry x:Name="e1" Center="140,100" RadiusX="100" RadiusY="60"/>
        </Path.Data>
    </Path>
    <Ellipse Fill="Green" Height="47" Name="ellipse1" Stroke="Black"
        VerticalAlignment="Top" Margin="86,96,321,0">
        <Ellipse.RenderTransform>
            <TransformGroup>
                <TranslateTransform X="-100" Y="-110"/>
                <MatrixTransform x:Name="mt1"/>
            </TransformGroup>
        </Ellipse.RenderTransform>
        <Ellipse.Triggers>
            <EventTrigger RoutedEvent="Page.Loaded">
                <BeginStoryboard>
                    <Storyboard x:Name="sb1" RepeatBehavior="Forever">
                        <MatrixAnimationUsingPath x:Name="ma1"
                            Storyboard.TargetName="mt1"
                            Storyboard.TargetProperty="Matrix"
                            Duration="0:0:10"/>
                    </Storyboard>
                </BeginStoryboard>
            </EventTrigger>
        </Ellipse.Triggers>
    </Ellipse>
    <Button Content="动画启动" Height="23" HorizontalAlignment="Left" Margin="0,202,0,0" Name="animationButton" VerticalAlignment="Top" Width="75" Click="animationButton_Click" />
</Grid>
</Window>
```

前台动画启动按钮的 animationButton_Click 事件的 CS 代码如下。

```csharp
private void animationButton_Click(object sender, RoutedEventArgs e)
{
    PathGeometry pg1 = new PathGeometry();
    pg1.AddGeometry(e1);
    ma1.PathGeometry = pg1;
    sb1.Begin(rectangle1);
}
```

运行完整的代码,当单击"动画启动"按钮后,矩形框沿椭圆的轨迹移动,页面显示在某一时刻的静态效果如图 8.7 所示。在 XAML 代码中用到了 Storyboard(故事板),Storyboard 可以视作 Animation 的容器,在其中可以包含任意类型的 Timeline 和动画。它的两个附加属性是 Storyboard.TargetName(故事板的目标名)和 Storyboard.TargetProperty(故事板的目标属性)。

再设置一个按钮沿曲线移动的路径动画,XAML 代码如下。

```
<!-- 保留 Window 代码部分 -->
<Canvas>
    <Canvas.Resources>
        <PathGeometry x:Key = "path" Figures = "M 10,100 C 35,0 135,0 160,100 180,190
          285,200 310,100" />
        <Storyboard x:Key = "pathStoryboard">
            <MatrixAnimationUsingPath PathGeometry = "{StaticResource path}"
              Storyboard.TargetName = "ButtonMatrixTransform"
                    Storyboard.TargetProperty = "Matrix"
                    DoesRotateWithTangent = "True"
                    Duration = "0:0:5" RepeatBehavior = "Forever">
            </MatrixAnimationUsingPath>
        </Storyboard>
    </Canvas.Resources>
    <Canvas.Triggers>
        <EventTrigger RoutedEvent = "Control.Loaded">
            <BeginStoryboard Storyboard = "{StaticResource pathStoryboard}" />
        </EventTrigger>
    </Canvas.Triggers>
    <Path Data = "{StaticResource path}" Stroke = "Black" StrokeThickness = "1" />
    <Button Width = "50" Height = "20">
        <Button.RenderTransform>
            <MatrixTransform x:Name = "ButtonMatrixTransform" />
        </Button.RenderTransform>
    </Button>
</Canvas>
</Window>
```

运行上述代码,按钮沿曲线移动,页面显示在某一时刻的静态效果,如图 8.8 所示。其中 Button 使用了 MatrixTransform,通过矩阵乘积修改坐标系统的元素绘制方式。

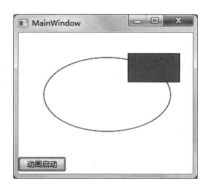

图 8.7 矩形沿椭圆轨迹移动的路径动画

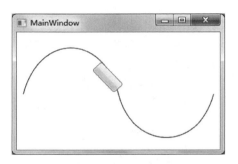

图 8.8 按钮沿曲线移动的路径动画

8.3 集成动画

在做开发时,经常会将动画集成到系统的其他部分。动画能处理任何值类型的任何属性,动画还可以集成到控件模板和文本类型中。

8.3.1 与控件模板集成

动画适合于构建媒体内容。从应用程序开发者的角度来看,能够把动画嵌入到控件中。

现在做一个具有边框的标签,当指针移到标签上,标签的边框加粗,边框颜色由原来的绿色变为天蓝色。

```
<Grid>
    <Label Width = "100" Height = "50" Content = "Welcome">
        <Label.Template>
            <ControlTemplate TargetType = "{x:Type Label}">
                <Border Name = "border1" CornerRadius = "4" BorderBrush = "Green"
                    BorderThickness = "2">
                    <ContentPresenter HorizontalAlignment = "Center" VerticalAlignment =
                        "Center" />
                </Border>
                <ControlTemplate.Triggers>
                    <EventTrigger RoutedEvent = "Label.MouseEnter">
                        <EventTrigger.Actions>
                            <BeginStoryboard>
                                <Storyboard>
                                    <ColorAnimation To = "SkyBlue"
                                        Storyboard.TargetName = "border1"
                                        Storyboard.TargetProperty = "BorderBrush.Color"/>
                                    <ThicknessAnimation To = "4"
                                        Storyboard.TargetName = "border1"
                                        Storyboard.TargetProperty = "BorderThickness"/>
                                </Storyboard>
                            </BeginStoryboard>
                        </EventTrigger.Actions>
                    </EventTrigger>
                </ControlTemplate.Triggers>
            </ControlTemplate>
        </Label.Template>
    </Label>
</Grid>
</Window>
```

运行上述代码,标签默认状态如图 8.9 所示。当鼠标指针移到标签上,页面效果如图 8.10 所示。

图 8.9 标签默认 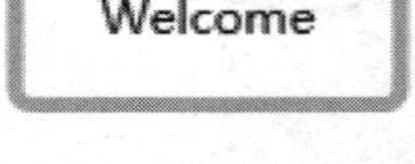　　　　　图 8.10 鼠标指针进入

从上述代码的结构可知，Storyboard 被定义为 ControlTemplate 的一部分。并设置 ControlTemplate.Triggers 属性，让其包含 EventTrigger 对象。当 BeginStoryboard 与动画板关联，动画则与模板化控件 Label 关联。

现在再往 ControlTemplate.Triggers 属性中添加一个 EventTrigger 对象，让其实现鼠标指针离开时，标签边框变细，边框颜色变成黑色的功能，页面效果如图 8.11 所示。新增 XAML 代码如下。

图 8.11　鼠标指针离开

```
<EventTrigger RoutedEvent = "Label.MouseLeave">
    <EventTrigger.Actions>
        <BeginStoryboard>
            <Storyboard>
                <ColorAnimation To = "Black" Storyboard.TargetName = "border1"
                                Storyboard.TargetProperty = "BorderBrush.Color"/>
                <ThicknessAnimation To = "1" Storyboard.TargetName = "border1"
                                Storyboard.TargetProperty = "BorderThickness"/>
            </Storyboard>
        </BeginStoryboard>
    </EventTrigger.Actions>
</EventTrigger>
```

8.3.2　与文本类型集成

在 8.3.1 节中，动画是集成到控件模板中使边框的精细与颜色发生变化，如何让动画对文本内容实施动画效果是本节的重点。

因为文本不是字符的任意堆砌，所以不能简单地把文本分割为以字符大小为标尺进行变换操作。每个 TextEffects 对象都包含 Transform、PositionStart 和 PositionCount 的实例，其决定着字符的效果。

实现对文本类型动画。动画效果形式为：每个字符以不同的时间移动，使用编程的方式来创建动画更容易。XAML 代码如下。

```
<Grid>
    <TextBlock FontSize = "24pt" Name = "textblock1" Foreground = "Green"> I am a textblock
            with animation
    </TextBlock>
</Grid>
```

后台 CS 代码如下。

```
public partial class MainWindow : Window
{
    public MainWindow()
    {
        InitializeComponent();
        Storyboard perChar = new Storyboard();
        textblock1.TextEffects = new TextEffectCollection();
        for (int i = 0; i < textblock1.Text.Length; i++)
        {
```

```
                TextEffect te = new TextEffect();
                te.Transform = new TranslateTransform();
                te.PositionStart = i;
                te.PositionCount = 1;
                textblock1.TextEffects.Add(te);
                DoubleAnimation da = new DoubleAnimation();
                da.To = 9;
                da.AccelerationRatio = .5;
                da.DecelerationRatio = .5;
                da.RepeatBehavior = RepeatBehavior.Forever;
                da.AutoReverse = true;
                da.Duration = TimeSpan.FromSeconds(2);
                da.BeginTime = TimeSpan.FromMilliseconds(250 * i);
                Storyboard.SetTargetProperty(da, new PropertyPath("TextEffects[" + i + "]
                .Transform.Y"));
                Storyboard.SetTargetName(da,textblock1.Name);
                perChar.Children.Add(da);
            }
            perChar.Begin(this);
        }
    }
```

运行完整的代码,动画效果如图 8.12 所示。分析代码的结构,因为要为每个字符创建效果,所以要读取文本的长度,用 for 循环实现这一迭代过程。在 for 循环中,构建 TextEffect 对象,把字符的起始点和数目关联到效果上,并且为动画创建一个变换。当然还需创建一个动画版,用于处理已创建好的多个 TextEffect 对象内容的变换。

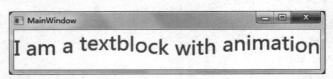

图 8.12　TextEffects 设置文本动画效果

8.4　媒　　体

在 WPF 中,对于媒体可以简单地定义为"与时间相关的数据流"。基于这个定义,前面所看到的动画,从技术上而言,都是媒体的一个片段。但是在本节中,讨论的重点是常见的外部媒体类型:音频和视频。音频文件是以时间为组织形式的波形内容;视频文件则是以时间组织的光栅图像。下面学习在 WPF 中如何播放音频与视频两类文件。

8.4.1　音频

WPF 中音频与视频文件都是通过 MediaTimeline。在任何音频和视频中,都存在时间线。当要播放媒体时,需要 MediaClock。

播放音频文件,需要如下 3 个步骤。

(1) 首先需要创建一个 MediaTimeline 的实例,并设置 Source 属性来指向媒体。

(2) 需要创建运行时钟。时钟对普通动画而言是必要的。但对于媒体文件,则需要 MediaPlayer 来管理媒体的状态。

(3) 为了给用户友好的提示,把 CurrentTimeInvalidated 事件挂接到时钟上,并更新 Window.Title 属性,用来显示音乐播放时间状态。

CS 代码如下。

```
public partial class MainWindow : Window
{
        MediaTimeline audioTL;
        MediaClock audioC;
        public MainWindow()
        {
            InitializeComponent();
            audioTL = new MediaTimeline();
            audioTL.Source = new Uri(@"f:\Try Everything.mp4");
            audioC = audioTL.CreateClock();
            MediaPlayer player = new MediaPlayer();
            player.Clock = audioC;
            audioC.CurrentTimeInvalidated += TimeChanged;
            audioC.Controller.Begin();
        }
        void TimeChanged(object sender, EventArgs e)
        {
            Title = audioC.CurrentTime.ToString();
        }
}
```

运行上述代码,页面显示效果如图 8.13 所示。代码中用到时间线、时钟和播放器 3 个对象,对于任何媒体播放的场景都是通用的。WPF 使用 MediaElement 类型隐藏了这些细节。

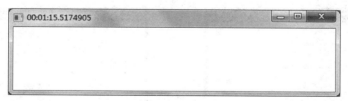

图 8.13　音乐播放状态

MediaElement 通过 Player 属性提供对播放器的访问,但对于高级的媒体动作,则需要在代码中创建时间线和时钟。如上面代码中用到的 CurrentTimeInvalidated 事件。

8.4.2　视频

使用上面提到的 MediaElement 来播放视频,XAML 代码如下。

```
<Grid>
        <MediaElement Source = "f:\功夫熊猫 3.mp4"/>
</Grid>
```

运行上述代码,视频播放页面显示效果如图 8.14 所示。

图 8.14 MediaElement 播放视频

8.5 小 结

动画是增强用户视觉体验的一种有效手段。本章从动画的基本概念出发,学习了动画的基本原理,对比了传统动画与 WPF 动画,介绍了 WPF 动画的三种类型和集成动画的方式,以及 WPF 中播放音频与视频两类文件的方法。

习题与实验 8

1. 简述动画的概念及原理。
2. 设计实现文字淡入、文字移动效果动画,如图 8.15 和图 8.16 所示。

图 8.15 文字淡入效果

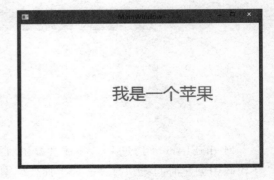

图 8.16 文字移动效果

3. 实现椭圆沿矩形轨迹移动的路径动画,由按钮控制开始,程序启动后,初始页面如图 8.17 所示。当单击"动画启动"按钮后,椭圆沿矩形轨迹移动的动画,页面显示在某一时刻的静态效果,如图 8.18 所示。

第 8 章 动画与媒体

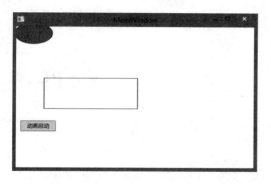
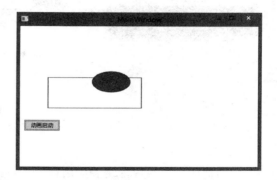

图 8.17　路径动画初始页面　　　　　　图 8.18　沿矩形轨迹移动动画

4. 实现小人快跑动画,如图 8.19 和图 8.20 所示。分析动画功能,则为两图中的两个图片的切换,还可以通过帧的进度控制跑步的速度。

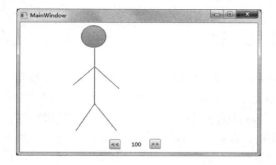

图 8.19　站立的小人页面　　　　　　图 8.20　运动的小人页面

第 9 章

动作

前面的学习内容围绕着应用程序使用控件构建及其不同的呈现方式,这些都属于平台之外的事物,本章将接触平台内部的东西。通常,当用户单击按钮、按下鼠标、移动鼠标、触摸屏幕等时,希望应用程序能以某种方式来响应这些动作。WPF 具有 3 个处理动作的常用方式,即事件、命令和触发器。WPF 将事件扩充到路由事件,在第 6 章已详细讲解过。本章重点介绍动作原则、命令系统及触发器的使用。

9.1 动作原则

WPF 动作的 3 个原则是元素合成、松散耦合和声明式动作。在第 1 章的 WPF 可视化树,显示了多个元素协同工作,这就要求动作采用元素合成的机制。如若事件的代码和事件处理过程的代码紧密耦合在一起,则控件能改变显示界面的行为会引发一些问题,所以松散耦合是合理的行为。最后,声明式的编程延伸到系统的各方面,故此,WPF 以声明的方式处理动作。

9.1.1 元素合成

在前面的学习中,了解到控件模型的 3 个原则:元素合成、富内容和简单的编程模型。

现在从简单的编程模型出发,用单击事件的监听代码,把处理过程附加到 Click 事件,代码如下。

```
Button b = new Button();
b.Content = "Click me";
b.Click += delegate { MessageBox.Show("You clicked me"); };
```

分析代码,按钮本身并没有被单击,被单击的是按钮显示界面元素,为了让这些元素协同工作,WPF 引入了 RoutedEvent(路由事件),路由事件可以穿越元素,对上面的代码进行修改,修改后的代码如下。

```
Button b = new Button();
b.Content = new Button();
b.Click += delegate { MessageBox.Show("You clicked me"); };
```

分析代码,第二个按钮作为第一个按钮的内容,此时,无论单击内部或外部的按钮都将引起事件的触发。元素合成的设计会影响到动作的处理及事件本身。

9.1.2 松散耦合

现在还以 Button 按钮为例,它支持直接的鼠标事件(MouseUp、MouseDown 等),也支持 Click 事件。Click 事件是比鼠标事件更高级的抽象。在按钮处于键盘焦点下,按下键盘上的 Enter 键时,Click 事件也会被触发。可以把鼠标事件看作物理(Physical)事件,Click 事件看作语义(Semantic)事件。

支持 Click 事件的控件除了 Button 以外,还有 CheckBox、RadioButton 和 HyperLink。只要这些控件被单击,都会触发 Click 事件。针对 Click 事件编写代码有两个好处:并不局限在某种特定的输入设备上(鼠标或键盘),也不是约束在按钮上。此时 Click 事件的代码编写只依赖于这个控件是否能被单击。把代码和动作产生过程进行解耦,处理过程更灵活。但是,尽管如此,事件本身还是受到耦合形式的约束,要求方法实现是某个特定的签名(Signature)。Button.Click 的委托定义形式如下。

```
public delegate void RoutedEventHandler(object sender,RoutedEventArgs e);
```

WPF 的动作耦合波谱如图 9.1 所示。它支持从物理事件(如 MouseUp)下的紧密耦合到完全语义提醒(如 ApplicationCommands.Close)的方式。

图 9.1 动作耦合波谱

由于松散耦合,可以编写模板,让控件产生意想不到的效果。

现在通过添加一个和 Close 命令(Command)挂接的按钮,为窗口编写一个模板,实现关闭窗口的功能。XAML 代码如下。

```xml
<ControlTemplate TargetType = "{x:Type Window}">
    <Grid>
        <StatusBar>
            <StatusBarItem>
                <Button
                    Command = "{x:Static ApplicationCommands.Close}">
                    Close
                </Button>
            </StatusBarItem>
        </StatusBar>
    </Grid>
</ControlTemplate>
```

在任何组件上引发了 Close 命令时,要通过命令绑定添加到窗口中让窗口关闭,后台 CS 代码如下。

```
public partial class MainWindow : Window
{
    public MainWindow()
```

```
        {
            InitializeComponent();
            CommandBindings.Add(new CommandBinding (ApplicationCommands.Close,CloseExecuted));
        }
        void CloseExecuted(object sender,ExecutedRoutedEventArgs e)
        {
            this.Close();
        }
}
```

命令在 WPF 中表示了最松散的耦合动作模型。在本例中,动作耦合提供了完全抽象的描述,可以改变窗口风格以使用完全不同的控件,而不会破坏任何动作的运转。

9.1.3 声明式动作

声明式编程的概念是 WPF 编程的基础。声明式逻辑为用户带来良好的体验,并为系统提供高级的服务。

处理动作的所有机制都支持这些原则。由于 WPF 路由事件已经在前面重点学习,因此下面需要深入了解命令系统。

9.2 命令系统

WPF 为人们准备了完善的命令系统,读者也许会有这样的疑问:"有了路由事件为什么还需要命令系统?"因为事件的作用是发布、传播消息,消息传达到了接收者,事件完成。至于如何响应事件送来的消息,事件并不做任何限制,每个接收者用自己的行为来响应事件。这就是说,事件不具有约束力,而命令具有约束力,这是两者的本质区别。

在实际编程工作中,只用事件而不用命令程序的逻辑一样被驱动得很好。以"保存事件的处理器"为例,程序员可以编写 Save()、SaveHandle() 和 SaveDocument() 等,这些都符合代码规范。但迟早有一天,整个项目会变得让人无法读懂,新来的程序员或修改 Bug 的程序员会受阻。如果使用命令,情况就会好很多。当 Save 命令到达某个组件时,命令会自动地去调用组件的 Save 方法。而这个方法可能定义在基类或接口中(即保证了这个方法是一定存在的),这样一来,约束了代码结构和命名规则。还可控制接收者"先做校验,再保存,最后退出"。也就是说命令除了可以约束代码,还可以约束步骤逻辑,让新来的程序员避免犯错,也让修改 Bug 的程序员有规律可循。

从上面的分析可知,命令可以看作一种逻辑约束行为,但这种逻辑约束行为可以被多种源调用,可以作用于多种目标上。如"复制""剪切"等命令,它们本身就是一种对剪贴板进行操作的逻辑约束行为。这些命令可以用在菜单项中,也可以在工具栏按钮上使用,还可以通过快捷键(Ctrl+C)来调用。由此可见,命令就像类(封装了多种信息),可以在多个地方调用。

下面学习命令系统的基本元素及元素间的关系。

9.2.1 基本元素及元素之间的关系

WPF 命令系统包含命令、命令源、命令目标和命令绑定 4 个基本元素,详细介绍如下。

- 命令(Command):要执行的动作,继承自 ICommand 接口的类,经常使用的有 RoutedCommand 类,但是,在实际使用中,一种有效的方法是在某个类中直接声明一个 RoutedCommand 类的成员字段,一般使用 Static 关键字,这样可以使得命令只与类有关,而不必理会其属于哪个实例。
- 命令源(CommandSource):命令的发送者,继承自 ICommandSource 接口的类。大部分界面的控件都实现了这个接口,如 Button、MenuItem 等。
- 命令目标(CommandTarget):命令的接收者,继承自 IInputElement 接口的类。例如,给文本框中粘贴内容,那么这个 TextBox 就是命令目标。
- 命令绑定(CommandBinding):将一些逻辑与命令绑定起来,如判断命令是否可以执行,以及执行完毕后做一些处理。

基本元素之间的关系体现在使用命令的过程中,使用命令可分为以下几个步骤。

(1) 创建命令类。它是获得一个实现 ICommand 接口的类的实例。

(2) 声明命令实例,使用命令时需要创建命令类的实例。

(3) 指定命令的源,指定由谁来发送这个命令。

(4) 指定命令目标。命令目标不是命令的属性,而是命令源的属性,是告知命令源向哪个组件发送命令,如果没有给命令源指定命令的目标,则系统会把当前拥有焦点的对象当作命令目标。

(5) 设置命令绑定。命令系统需要 CommandBinding 在执行命令前不断地问询是不是可以执行,并且在命令执行完毕给出反馈。

命令目标与命令绑定间的关系是,当某个 UI 控件被命令源确定为目标后,命令源就会不停地向命令目标发送可路由的 PreviewCanExecute 和 CanExecute 附加事件,事件会沿着 UI 逻辑树向上传递并被命令绑定获知,命令绑定并将获知的消息实时报告给命令。与此相似,如果命令被发送出来并到达命令目标,命令目标则会发送 PreviewExecuted 和 Executed 两个附加事件。这两个事件也会沿着 UI 逻辑树向上传递并被命令绑定所获知,对于那些与业务逻辑无关的通用命令,这些后续任务才是最重要的。

上述 WPF 命令描述过程,可知命令系统的基本元素关系图,如图 9.2 所示。

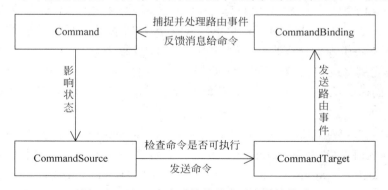

图 9.2 WPF 命令系统的基本元素间的关系

现在定义编辑(Edit)菜单,其下包含复制(Copy)、剪切(Cut)和粘贴(Paste)3 个菜单项,这些菜单项对菜单下的文本框(TextBox)内容进行编辑操作。菜单项通过 WPF 命令的形式实现，XAML 代码如下。

```xml
<Grid>
    <Grid.RowDefinitions>
        <RowDefinition Height = "23" />
        <RowDefinition />
    </Grid.RowDefinitions>
    <Menu Grid.Row = "0" Grid.Column = "0">
        <MenuItem Header = "Edit">
            <MenuItem x:Name = "menuCopy" Header = "Copy"
                      Command = "ApplicationCommands.Copy"
                      CommandTarget = "{Binding ElementName = textBox1}"/>
            <MenuItem x:Name = "menuCut" Header = "Cut"
                      Command = "ApplicationCommands.Cut" />
            <MenuItem x:Name = "menuPaste" Header = "Paste"
                      Command = "ApplicationCommands.Paste" />
        </MenuItem>
    </Menu>
    <TextBox Grid.Row = "1" Grid.Column = "0" x:Name = "textBox1"
             TextWrapping = "Wrap" AcceptsReturn = "True" />
</Grid>
```

运行上述代码，在 textbox1 中键入信息后，单击 Edit 菜单，页面显示效果如图 9.3 所示。

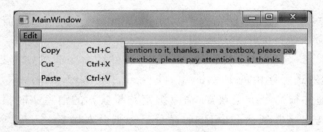

图 9.3　WPF 命令实现 Edit 菜单

分析上述代码，与命令系统中的基本元素相对应，此例中 ApplicationCommands.Copy、ApplicationCommands.Cut 和 ApplicationCommands.Paste 是命令，这些命令赋值给了菜单项的 Command 属性，实现了 ICommandSource 接口的元素都拥有该属性，其 Command 属性就指示了其将引发的命令；3 个 MenuItem 控件是命令源；textbox1 文本框就是命令目标；命令绑定到系统定义，对于文本框的"复制""剪切"和"粘贴"操作。

在 WPF 的命令系统中的 4 个基本元素都派生自不同的类，它们所在的类之间的调用关系如图 9.4 所示。

由图 9.4 可知，WPF 命令系统中的命令绑定（CommandBinding）是将一个命令与实现该命令的事件处理程序关联。CommandBinding 类包含一个 Command 属性以及 Executed、PreviewExecuted、PreviewCanExecute 和 CanExecute 事件。Command 是 CommandBinding 要与之关联的命令。附加到 PreviewExecuted 和 Executed 事件的事件处理程序实现命令逻辑。附加到 PreviewCanExecute 和 CanExecute 事件的事件处理程序确定命令是否可以在当前命令目标上执行。CanExecute 事件和 PreviewCanExecute 事件，通过其 EventArgs 参数中的 CanExecute 属性，设置其命令是否可以执行，并且系统会自动地与命令目标的某些特定

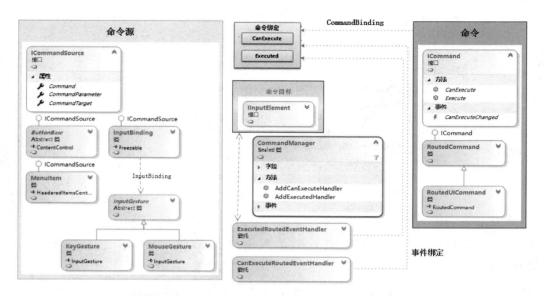

图 9.4　WPF 命令系统的基本元素派生类之间的调用关系

属性进行绑定，如 Button、MenuItem 等在 CanExecute 属性的值设为 False 时，会"灰化"，不可用。Executed 事件和 PreviewExecuted 事件的代码，是执行命令的真正代码。

对上面的菜单，再加一个文件菜单，将 ApplicationCommands.Save 绑定到菜单栏的 Save 菜单项中，当文本框中没有文本时不可用。实现文件菜单，需添加 Window 命令绑定的 XAML 代码如下。

```xml
<Window.InputBindings>
        <KeyBinding Command = "ApplicationCommands.Save"/>
</Window.InputBindings>
<Window.CommandBindings>
        <CommandBinding Command = "ApplicationCommands.Save"
                        CanExecute = "CommandBinding_Save_CanExecute"
                        Executed = "CommandBinding_Save_Executed" />
</Window.CommandBindings>
```

编写 File 菜单下的 Save 菜单项的 XAML 代码如下。

```xml
<MenuItem Header = "File">
            <MenuItem x:Name = "menuSave" Header = "Save"
                      Command = "ApplicationCommands.Save" />
</MenuItem>
```

后台实现 Save 命令的事件处理程序的 CS 代码如下。

```csharp
private void CommandBinding_Save_CanExecute(object sender, CanExecuteRoutedEventArgs e)
{
            if (textBox1.Text == string.Empty)
            {   //如果文本框中没有任何文本,则不可以保存
                e.CanExecute = false;
            }
```

```
        else
        {
            e.CanExecute = true;
        }
    }
    private void CommandBinding_Save_Executed(object sender, ExecutedRoutedEventArgs e)
    {   // 保存文件对话框
        SaveFileDialog save = new SaveFileDialog();
        save.Filter = "文本文件 t|*.txt|所有文件 t|*.*";
        bool? result = save.ShowDialog();
        if (result.Value)
        {
            // 执行保存文件操作
        }
    }
```

运行完整的代码,若文本框中无文本,单击 Save 菜单项后,命令不可用,页面显示效果如图 9.5 所示。

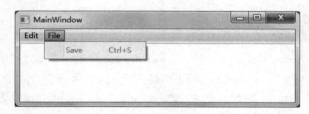

图 9.5　WPF 命令实现文件菜单的不可用状态

文本框中输入文本后,单击 File 菜单后,Save 菜单项可用,页面显示效果如图 9.6 所示。

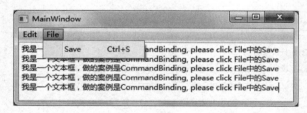

图 9.6　WPF 命令实现文件菜单的可用状态

9.2.2　ICommand 接口

WPF 命令系统的核心是 System.Windows.Input.ICommand 接口,而且 WPF 中的命令都继承自 ICommand 接口。该接口定义了命令的工作原理,它包含两个方法和一个事件,定义如下。

```
public interface ICommand
{
    void Executed(object parameter);
    bool CanExecute(object parameter);
    event EventHandler CanExecuteChanged;
}
```

还有一些类继承自 ICommandSource，这些类包括 ButtonBase、MenuItem、Hyperlink 和 InputBinding。而 Button、GridViewColumnHeader、ToggleButton 和 RepeatButton 继承自 ButtonBase。System.Windows.Input.KeyBinding 和 MouseBinding 继承自 InputBinding。

9.2.3　RoutedCommand 类

创建命令时，不会直接实现 ICommand 接口，而是使用 System.Windows.Input.RoutedCommand 类，该类自动实现 ICommand 接口。RoutedCommand 类是 WPF 中唯一实现了 ICommand 接口的类。所有 WPF 命令都是 RoutedCommand 类（及其派生类）的实例。

WPF 命令系统中的一个重要概念 RoutedCommand 类不包含任何应用程序逻辑。它只是代表一个命令。这意味着各个 RoutedCommand 对象具有相同的功能。

RoutedCommand 类为事件冒泡和隧道添加了一些额外的基础结构。鉴于 ICommand 接口封装了命令的思想，可以被触发的动作并且可以被启用或禁用——RoutedCommand 类对命令进行了修改，从而使命令可以在 WPF 元素层次结构中冒泡以便获得正确的事件处理程序。

9.2.4　RoutedUICommand 类

在程序中处理的大部分命令不是 RoutedCommand 对象，而是 RoutedUICommand 类的实例，RoutedUICommand 类继承自 RoutedCommand 类。事实上，WPF 提供的所有预先构建好的命令都是 RoutedUICommand 对象。

RoutedUICommand 类用于创建具有文本属性的命令，这些文本显示在用户界面中的某些地方（如菜单项文本、工具栏按钮的工具提示）。RoutedUICommand 类只增加了一个 Text 属性，它是为命令显示的文本。

为命令定义命令文本，而不是直接在控件上定义文本的优点是可在一个位置执行本地化。但是如果命令文本永远不会在用户界面上的任何地方显示，此时 RoutedUICommand 类和 RoutedCommand 类是等效的。

9.2.5　WPF 命令库

WPF 中提供了一些便捷的命令库，它们是 ApplicationCommands、MediaCommands、ComponentCommands、NavigationCommands 和 EditingCommands。下面给出 WPF 命令库的类及命令值，如表 9.1 所示。

表 9.1　命令库及命令值

命　令　类	示　　例
ApplicationCommands	Close、Cut、Copy、Paste、Save、Print
NavigationCommands	BrowseForward、BrowseBack、Zoom、Search
EditingCommands	AlignXXX、MoveXXX、SelectXXX
MediaCommands	Play、Pause、NextTrack、IncreaseVolume、Record、Stop
ComponentCommands	MoveXXX、SelectXXX、ScrollXXX、ExtendSelectionXXX

表9.1中的XXX代表操作的集合,如MoveNext和MovePrevious。其中,ApplicationCommands为默认的命令类,引用其中的命令时可以省略ApplicationCommands。

9.2.6 命令与数据绑定

WPF中的大部分事件都和每个控件的实现紧密关联,用户在自定义命令时,通常都会使用事件。

现在以关闭程序为例,首先在File菜单中的菜单项关闭程序,使用XAML的代码如下。

```
< MenuItem Header = "File">
    < MenuItem x:Name = "menuExit" Header = "Exit" Click = "ExitClicked"/>
</MenuItem>
```

后台实现关闭命令的事件处理程序的CS代码如下。

```
private void ExitClicked(object sender, RoutedEventArgs e)
{
    Application.Current.Shutdown();
}
```

下面添加退出应用程序的超级链接控件到文本块(TextBlock),用TextBlock控件替换原代码中的TextBox控件,XAML代码如下。

```
< TextBlock Grid.Row = "1" Grid.Column = "0" > Welcome to my program. If you fell bored,
            you can < Hyperlink Click = "ExitClicked"> exit </Hyperlink>
</TextBlock>
```

将关闭程序代码添加到前面的案例中,运行完整的代码程序,页面显示效果如图9.7所示。ExitClicked的签名和Hyperlink.Click事件是兼容的,它们都是执行退出应用程序的命令。另外,可以把命令的事件处理程序添加到XAML代码中,所以图形设计器在为应用程序构建UI时,无须知道其续写到哪里。

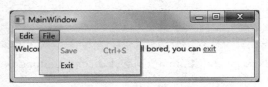

图9.7 事件处理程序与超链接实现退出命令

现在在菜单上放置"重做"和"撤销"两个菜单,XAML代码如下。

```
< MenuItem Command = "ApplicationCommands.Redo" Header = "{Binding
        Path = Command.Text, RelativeSource = {RelativeSource Self}}"/>
< MenuItem Command = "ApplicationCommands.Undo" Header = "{Binding
        Path = Command.Text, RelativeSource = {RelativeSource Self}}"/>
```

运行上述代码,页面显示效果如图9.8所示。其中,"Header = "{Binding Path = Command.Text,…}""代码中的菜单文本绑定到了命令的Text属性。因为,当一个命令为RoutedUICommand类型,那么该命令将有一个Text属性来说明该命令对应到的文本名称。该 Text 属性会自动本地化的，由于当前的计算机使用语言是简体中文的，故该菜单项显示的是"重做、撤销"，如果计算机使用的语言是英语，菜单项显示的将是"Redo、Undo"。

图 9.8　命令绑定实现"重做"和"撤销"菜单

由于 Command 和 CommandParameter 都是元素上的属性，因此它能够被设置成数据，这就使得命令和数据绑定集成到一起。使用命令也可以实现数据驱动的逻辑。

现在在上面案例的基础上，添加一个列表框和用于显示每个文件名的数据模板，XAML 代码如下。

```
<ListBox Name = "listBox1" Grid.Row = "3">
    <ListBox.ItemTemplate>
        <DataTemplate>
            <TextBlock Text = "{Binding Path = Name}"/>
        </DataTemplate>
    </ListBox.ItemTemplate>
</ListBox>
```

在后台，将 ItemsSource 属性设置为文件的列表，CS 代码如下。

```
public MainWindow()
{
    InitializeComponent();
    FileInfo[] fileList = new DirectoryInfo("c:\\").GetFiles("*.*");
    listBox1.ItemsSource = fileList;
}
```

运行完整的代码，页面显示效果如图 9.9 所示。

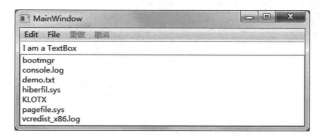

图 9.9　显示文件列表

下面添加一个用来显示文件的按钮，并且只显示文本文件。所以，需要在加载的文件上提供某种类型的过滤器，以实现 Blocked 和 Open 两个命令，后台的 CS 代码如下。

```
public partial class MainWindow : Window
{
```

```csharp
        public static readonly RoutedCommand BlockedCommand = new RoutedCommand("Blocked",
typeof(MainWindow));
        public static readonly RoutedCommand OpenCommand = new RoutedCommand("Open",
typeof(MainWindow));
    ...
    }
```

再为命令编写处理过程,CS 代码如下。

```csharp
using System.Diagnostics;
...
public MainWindow()
{
        InitializeComponent();
        CommandBindings.Add(new
CommandBinding(BlockedCommand,delegate(object sender,
            ExecutedRoutedEventArgs e)
            {
                MessageBox.Show((string)e.Parameter, "Blocked");
            }));
        CommandBindings.Add(new CommandBinding(OpenCommand,
            delegate(object sender, ExecutedRoutedEventArgs e)
            {
                Process.Start("notepad.exe", (string)e.Parameter);
            }));
}
```

在对文件操作时,某些文件用 OpenCommand 打开,而某些文件用 BlockedCommand 锁定,所以使用 IValueConverter 把文件名转换为 ICommand。编写文件转换成命令的转换器,检查文件扩展名功能,后台 CS 代码如下。

```csharp
using System.Windows.Data;
using System.Globalization;
using System.IO;
...
public class FileToCommandConverter : IValueConverter
{
    public object Convert(object value, Type targetType, object parameter, CultureInfo culture)
    {
        string extxt = ((FileInfo)value).Extension.ToLowerInvariant();
        if (extxt == ".txt")
        {
            return MainWindow.OpenCommand;
        }
        return MainWindow.BlockedCommand;
    }
    public object ConvertBack(object value, Type targetType, object parameter, CultureInfo culture)
    {
        return value;
```

    }
}

下面在文件数据模板中放置按钮。在命令参数(文件名)中使用数据绑定,XAML 代码如下。

```
<DataTemplate>
    <WrapPanel>
        <TextBlock Text = "{Binding Path = Name}"/>
        <Button CommandParameter = "{Binding Path = FullName}">
            <Button.Command>
                <Binding>
                    <Binding.Converter>
                        <local:FileToCommandConverter />
                    </Binding.Converter>
                </Binding>
            </Button.Command>
            Display
        </Button>
    </WrapPanel>
</DataTemplate>
```

运行上述程序,页面显示效果如图 9.10 所示。只有.txt 文件可以显示其内容。命令允许在 UI 和行为之间实现松散耦合,这就让应用程序行为的定义过程也基于数据驱动的方式。有一些并非应用程序逻辑的行为,它是用来控制显示状态的。例如,当用户在按钮上移动鼠标指针时,按钮高亮显示。这样的显示逻辑可以通过命令或事件实现。当这个行为用代码来实现时,此时显示和行为之间又回到了紧耦合。下面通过触发器来解决这个问题。

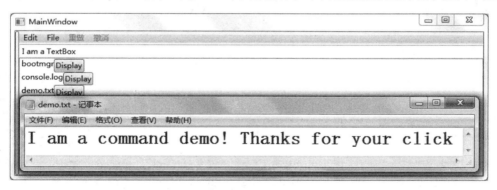

图 9.10 数据绑定命令

9.3 触 发 器

在第 8 章中,触发器可以声明式地将动画的启动和控件模板关联到一起。触发器有 Trigger、DataTrigger 和 EventTrigger 3 种类型。其中,Trigger 是属性触发器,它是当 DependencyProperty 的值发生改变时触发;DataTrigger 是数据触发器,是当普通.NET 属性的值发生改变时触发;EventTrigger 是事件触发器,当路由事件被触发时调用。另外,还有 MultiTriggers 和 MultiDataTriggers 两种集合触发器类型。

9.3.1 数据触发器

数据触发器（DataTrigger）只能被用在数据模板中，它以声明的方式设定数据模型值的动作。这样看来，DataTrigger 就是一个在标记中的简单值转换器，它使用数据绑定方式从数据模型中取值，并在值匹配时，可调用 Setter 和 EventSetter。

在开始讲解本节案例之前，先在后台创建一个名为 DataTriggerDemo 的类，便于后期操作。

```
public class DataTriggersDemo
{
    public static readonly RoutedCommand BlockedCommand = new RoutedCommand("Blocked", typeof(MainWindow));
    public static readonly RoutedCommand OpenCommand = new RoutedCommand("Open", typeof(MainWindow));
}
```

继续使用上节命令与数据绑定案例，使用 DataTrigger 替换值转换器，把 Command 属性设置成 BlockedCommand，XAML 代码如下。

```
<DataTemplate>
    <WrapPanel>
        <TextBlock Text="{Binding Path=Name}"/>
        <Button x:Name="_displayButton" Content="Display"
            Command="{x:Static local:DataTriggersDemo.BlockedCommand}"
            CommandParameter="{Binding Path=FullName}"/>
    </WrapPanel>
</DataTemplate>
```

在文件扩展名为".txt"时，设置数据触发器，新增 XAML 代码如下。

```
<DataTemplate>
    <WrapPanel>
        ...
    </WrapPanel>
    <DataTemplate.Triggers>
        <DataTrigger Binding="{Binding Path=Extension}" Value=".txt"/>
    </DataTemplate.Triggers>
</DataTemplate>
```

当扩展名为".txt"时，就把 Command 属性设置为 OpenCommand。此时，需要添加一个 Setter（设置器）到 DataTrigger。为设置器赋值的 XAML 代码如下。

```
<DataTemplate>
    <WrapPanel>
        ...
    </WrapPanel>
    <DataTemplate.Triggers>
        <DataTrigger Binding="{Binding Path=Extension}" Value=".txt">
            <Setter TargetName="_displayButton" Property="Command"
                Value="{x:Static local:DataTriggersDemo.OpenCommand}"/>
        </DataTrigger>
    </DataTemplate.Triggers>
</DataTemplate>
```

因为 DataTrigger 支持一组 Setter 对象,所以可以执行多个动作来响应数据值。在此,添加第二个设置器来显示可执行的命令。完整的 XAML 代码如下。

```
<DataTemplate>
    <WrapPanel>
        <TextBlock Text = "{Binding Path = Name}"/>
        <Button x:Name = "_displayButton" Content = "Block"
            Command = "{x:Static local:DataTriggersDemo.BlockedCommand}"
            CommandParameter = "{Binding Path = FullName}"/>
    </WrapPanel>
    <DataTemplate.Triggers>
        <DataTrigger Binding = "{Binding Path = Extension}" Value = ".txt">
            <Setter TargetName = "_displayButton" Property = "Command"
                Value = "{x:Static local:DataTriggersDemo.OpenCommand}"/>
            <Setter TargetName = "_displayButton" Property = "Content"
                Value = "Open"/>
        </DataTrigger>
    </DataTemplate.Triggers>
</DataTemplate>
```

运行上述代码,页面显示效果如图 9.11 所示。使用 Setter 设置了两组属性,当文件名后缀是".txt"时,按钮的 Content 为 Open;否则,按钮的 Content 为 Block。

图 9.11 DataTrigger 设置多个属性

使用 DataTrigger 时,属性设置数据的值与所设定的值类型完全匹配。由上节案例可知,值转换器调用 ToLowerInvariant 来处理不同的文件名。为了处理不同的文件名,创建一个简单的值转换器,后台 CS 代码如下。

```
using System.Windows.Data;
using System.Globalization;
…
public class ToLowerInvariantConverter:IValueConverter
{
     public object Convert(object value, Type targetType, object parameter, CultureInfo culture)
    {
        return ((string)value).ToLowerInvariant();
    }
    public object ConvertBack(object value, Type targetType, object parameter, CultureInfo culture)
    {
        return value;
    }
}
```

然后再把这个转制器附加到触发器上,XAML 代码如下。

```
<DataTemplate.Triggers>
    <DataTrigger Value = ".txt">
        <DataTrigger.Binding>
            <Binding Path = "Extension">
                <Binding.Converter>
```

```
                    <local:ToLowerInvariantConverter />
                </Binding.Converter>
            </Binding>
        </DataTrigger.Binding>
        <Setter TargetName = "_displayButton" Property = "Command"
                Value = "{x:Static local:DataTriggersDemo.OpenCommand}"/>
        <Setter TargetName = "_displayButton" Property = "Content"
                Value = "Open"/>
    </DataTrigger>
</DataTemplate.Triggers>
```

使用 DataTrigger，把所有 UI 独立的逻辑（命令绑定）移动到标记中，并简化转换器。这种方法将主要的逻辑放置在标记中，显示界面具有了创建工具的能力，让 UI 和应用程序逻辑分离。

9.3.2 属性触发器

属性触发器（Trigger）能被用于 Style、ControlTemplate、DataTemplate。也就是说，其适用于控件模板或样式。

下面在 Style 中使用 Trigger，用 DockPanel 布局，内部放置一个 TextBox。当 TextBox 的 Text 属性值发生改变时，引发触发器。XAML 代码如下。

```
<DockPanel>
    <TextBox TextWrapping = "Wrap" Margin = "5" DockPanel.Dock = "Top">
        <TextBox.Style>
            <Style TargetType = "TextBox">
                <Style.Triggers>
                    <Trigger Property = "Text" Value = "hello, everyone!">
                        <Setter Property = "Background" Value = "Pink"/>
                    </Trigger>
                </Style.Triggers>
            </Style>
        </TextBox.Style>
    </TextBox>
</DockPanel>
```

运行上述代码，在 TextBox 输入"hello，everyone！"，它的 Background（背景）变为粉红色，如图 9.12 所示。

下面在 TextBox 控件下加入 CheckBox 控件，实现当鼠标指针滑过 CheckBox 时，其前景色变为红色，XAML 代码如下。

```
<CheckBox Content = "Style Trigger MouseOver Red">
    <CheckBox.Resources>
        <Style TargetType = "{x:Type CheckBox}">
            <Setter Property = "Foreground" Value = "SkyBlue"/>
            <Style.Triggers>
                <Trigger Property = "IsMouseOver" Value = "True">
                    <Setter Property = "Foreground" Value = "Red"/>
                </Trigger>
            </Style.Triggers>
        </Style>
    </CheckBox.Resources>
```

```
</CheckBox>
```

运行上述代码,页面显示效果如图 9.13 所示。当鼠标指针滑过时,字体变成红色 Trigger。

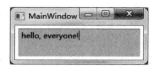

图 9.12　属性触发器

图 9.13　鼠标滑过前景色变红

9.3.3　多条件触发器

截至目前,数据触发器和属性触发器都是针对单个条件,也就是说当某一个条件满足时就会触发。而现实中,人们可能需要满足很多个条件时才触发一系列操作,这时就需要用到 MultiTrigger 或 MultiDataTrigger。它们都具有一个 Conditions 集合用来存放一些触发条件,这里的 Condition 之间是 And 关系,当所有条件都满足时,Setter 集合才会被调用。根据名称可知:MultiTrigger 用于实现多个属性(这里的属性指依赖属性)同时满足条件时调用;MultiDataTrigger 用于实现多个数据触发器(这里的属性指.NET 属性)同时满足条件时调用。

下面使用 MultiTrigger 可以实现多条件触发,XAML 代码如下。

```
<Window.Resources>
    <Style TargetType = "CheckBox">
        <Style.Triggers>
            <MultiTrigger>
                <!-- 条件列表 -->
                <MultiTrigger.Conditions>
                    <Condition Property = "IsChecked"
                               Value = "true" />
                    <Condition Property = "Content"
                               Value = "浪花淘尽英雄" />
                </MultiTrigger.Conditions>
                <MultiTrigger.Setters>
                    <Setter Property = "FontSize" Value = "35" />
                    <Setter Property = "Foreground" Value = "Red" />
                </MultiTrigger.Setters>
            </MultiTrigger>
        </Style.Triggers>
    </Style>
</Window.Resources>
<StackPanel>
    <CheckBox Content = "滚滚长江东逝水" Margin = "5" />
    <CheckBox Content = "浪花淘尽英雄" Margin = "5,0" />
    <CheckBox Content = "青山依旧在" Margin = "5" />
    <CheckBox Content = "几度夕阳红" Margin = "5,0" Width = "496" />
</StackPanel>
</Window>
```

运行上述代码,页面显示效果如图 9.14 所示。由代码中的条件列表可知,当选中 CheckBox 的 IsChecked 属性为真,且 Content 的值为"浪花淘尽英雄"时,MultiTrigger 触发,此时,符合条件的 CheckBox 的 FontSize 属性值 35;它的 Foreground 是 Red。

由于 MultiDataTrigger 是用于实现多个.NET 属性同时满足条件时才调用。接下来,先创建 User 类,包含姓名和年龄两个属性,CS 代码如下。

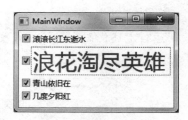

图 9.14 MultiTrigger

```
public class User
{
    string name;
    public string Name
    {
        get { return this.name; }
        set { this.name = value; }
    }
    int age;
    public int Age
    {
        get { return this.age; }
        set { this.age = value; }
    }
    public User() { }
    public User(string name, int age)
    {
        this.name = name;
        this.age = age;
    }
}
```

在< Window.Resources ></Window.Resources >中,构造 MultiDataTrigger 的条件,XAML 代码如下。

```
xmlns:local = "clr-namespace:MultDataTriggerDemo"
Title = "MainWindow" Height = "350" Width = "525" Loaded = "Window_Loaded">
<Window.Resources>
    <Style TargetType = "Button">
        <Style.Triggers>
            <MultiDataTrigger>
                <!-- 条件列表 -->
                <MultiDataTrigger.Conditions>
                    <Condition Binding = "{Binding Path = Name}" Value = "梦想成真"/>
                    <Condition Binding = "{Binding Path = Age}" Value = "89"/>
                </MultiDataTrigger.Conditions>
                <Setter Property = "FontSize" Value = "18" />
                <Setter Property = "Foreground" Value = "Red"/>
            </MultiDataTrigger>
        </Style.Triggers>
```

```
        </Style >
</Window.Resources >
```

用 Grid 布局前台页面，Grid 共 3 行 2 列，第 0 行第 0 列放置一个 TextBlock；第 0 行第 1 列放置一个 TextBox；第 1 行第 0 列放置一个 TextBlock，第 1 行第 1 列放置一个 TextBox；第 2 行放置一个 Button 让其占据 2 列。XAML 代码如下。

```
< Grid x:Name = "grid1">
    < Grid.RowDefinitions >
        < RowDefinition/>
        < RowDefinition/>
        < RowDefinition/>
    </Grid.RowDefinitions >
    < Grid.ColumnDefinitions >
        < ColumnDefinition/>
        < ColumnDefinition/>
    </Grid.ColumnDefinitions >
    < TextBlock Height = "23" Text = "Name:" HorizontalAlignment = "Left"
            Margin = "12,12,0,0" VerticalAlignment = "Center" />
    < TextBlock Height = "23" Grid.Row = "1" Grid.Column = "0" Margin = "12,12,0,0"
            HorizontalAlignment = "Left"Text = "Age:"VerticalAlignment = "Center"/>
    < TextBox Text = "{Binding Name}" HorizontalAlignment = "Center" Width = "120"
            Grid.Column = "1" Margin = "0,12,12,0" Name = "nameTextBox" VerticalAlignment =
            "Center" />
    < TextBox Text = "{Binding Age}" Grid.Row = "1" Grid.Column = "1"
            HorizontalAlignment = "Center" Margin = "0,12,12,0" Width = "120"
            Name = "ageTextBox" VerticalAlignment = "Center"/>
    < Button Content = "{Binding Path = Name}" Name = "button1" Grid.Row = "2"
            Grid.ColumnSpan = "2" Margin = "12"/>
</Grid >
```

在后台定义一个 User 类的实例，并指定数据上下文。CS 代码如下。

```
public partial class MainWindow : Window
{
        public MainWindow()
        {
                InitializeComponent();
        }
        private void Window_Loaded(object sender, RoutedEventArgs e)
        {
                User user1 = new User();
                grid1.DataContext = user1;
        }
}
```

运行完整的代码，在 Name 与 Age 所绑定的 TextBox 控件中分别输入"梦想成真"和"89"，再单击 Button 后，页面显示效果如图 9.15 所示。

在本节中，各种触发器都设置在 Style 内部。故从某种意义上讲，触发器可以看作一种 Style，因为它包含

图 9.15 MultiDataTrigger

有一个 Setter 集合,并根据一个或多个条件执行 Setter 中的属性改变。Styles 是放置触发器的最好位置。但对于每个 FrameworkElement 来说,都有 Triggers 集合,也可以放在 Triggers 集合中。

9.4 小　　结

本章介绍了动作原则、命令系统及触发器的使用。其中,触发器适用于模板或样式中。Trigger 和 EventTrigger 能被用于控件模板或样式中,DataTrigger 只能被用于数据模板中。

习题与实验 9

1. 简述路由事件与命令的区别,并从生活中寻找相关案例。
2. 设计 Windows 的记事本,实现复制、粘贴、剪切、保存、退出等功能。
3. 根据本章多条件触发器的相关知识,分别实现 MultiTrigger 和 MultiDataTrigger 两个案例,并解释为什么只有"浪花淘尽英雄"样式发生变化。

第 10 章

资源

在第 1 章中第一次引入资源的概念，由其案例可见，定义资源以后，可以在多处重复利用。本章将开始深入地学习资源。通过资源的概念学习，认识资源的常见类型，学会如何引用资源，进一步创建及使用资源字典。并简单触及用 ResourceDictionary 来管理多个 Resources 文件（这是换肤的基础）。

10.1 资源概述

WPF 的资源用于保存可以被重复利用的样式、对象定义及传统的资源（如二进制数据、图片等）。使用 WPF 的资源在应用程序中实现外形的更换比以往更简单。

10.1.1 资源的定义

资源是保存在可执行文件中的一种不可执行数据。在 WPF 的资源中，几乎可以包含图像、字符串等所有的任意 CLR 对象，只要对象有一个默认的构造函数和独立的属性。也就是说，应用程序中非程序代码的内容，如点阵图、颜色、字型、动画、影片文档及字符串常量值，可将它们从程序中独立出来，单独包装成"资源（Resource）"。

下面在 WPF 资源中，定义一种复用的 SolidColorBrush 对象，再让按钮和文本框的背景及矩形的填充颜色均使用 SolidColorBrush 对象。

```
<!-- 保留 Window 代码部分 -->
    <Window.Resources>
        <SolidColorBrush x:Key = "goldBrush" Color = "Gold"/>
    </Window.Resources>
    <StackPanel>
        <Button Margin = "5" Content = "Button with goldBrush"
         Background = "{StaticResource goldBrush}"/>
        <TextBlock Margin = "5" Text = "This is a TextBlock"
         Background = "{StaticResource goldBrush}" />
        <Rectangle Margin = " 5 " Width = " 500 " Height = " 25 " Fill = " {StaticResource
         goldBrush}"/>
    </StackPanel>
</Window>
```

运行上述代码，页面显示效果如图 10.1 所示。

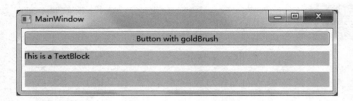

图 10.1 资源定义及引用

10.1.2 资源可用范围

WPF 具有封装和存取资源机制,开发人员可将资源建立在应用程序的不同范围上。WPF 中,资源定义的位置决定了该资源的可用范围。通常资源可以定义在控件、窗体、应用程序和资源字典中。对资源的可用范围,详细介绍如下。

(1) 控件(对象)级:资源定义在某个 ContentControl 中,作为其子容器、子控件共享的资源。

现在定义 Button 上的资源,TextBlock 是 Button 的子控件,调用 Button 资源的 XAML 代码如下。

```
<!-- 保留 Window 代码部分 -->
<StackPanel>
    <Button>
        <Button.Resources>
            <SolidColorBrush x:Key = "myYellowBrush" Color = "Yellow" />
        </Button.Resources>
        <Button.Content>
            <TextBlock Text = " I am a Button.Content "
              Background = "{StaticResource myYellowBrush}" />
        </Button.Content>
    </Button>
</StackPanel>
</Window>
```

运行上述代码,页面显示效果如图 10.2 所示,Button 的子控件 TextBlock,共享了 Button 上定义的 SolidColorBrush 对象资源。

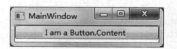

图 10.2 控件级资源

(2) 窗体级:资源定义在 Window 或 Page 层级的 XAML 文档中,本窗体或页面的所有的控件都可使用,在资源定义中的示例,则属于窗体级。

(3) 应用程序级:资源定义在 App.xaml 中,资源可用到应用程序内的任何地方。下面在 App.xaml 中定义 SolidColorBrush,XAML 代码如下。

```
<Application x:Class = "ApplicationResource0.App"
        xmlns = "http://schemas.microsoft.com/winfx/2006/xaml/presentation"
        xmlns:x = "http://schemas.microsoft.com/winfx/2006/xaml"
        StartupUri = "MainWindow.xaml">
    <Application.Resources>
```

```
        < SolidColorBrush x:Key = "myGoldBrush" Color = "Gold" />
    </Application.Resources >
</Application >
```

在 MainWindow.xaml 中引用资源,XAML 代码如下。

```
<!-- 保留 Window 代码部分 -->
    < StackPanel >
        < Button Margin = "5" Background = "{StaticResource myGoldBrush}"> I am a Button
        </Button>
    </StackPanel >
</Window >
```

运行完整的代码,页面显示效果如图 10.3 所示,在 MainWindow.xaml 文档中,Button 的属性 Background 引用了 SolidColorBrush 资源。

图 10.3 应用程序级资源

(4) 字典级:资源封装成一个资源字典,定义到一个 ResourceDictionary 的 XAML 文件中,还可被其他应用程序重复使用,将在 10.4 节中重点介绍。

每一个框架级元素(FrameworkElement 或者 FrameworkContentElement)都有一个资源属性。每一个在资源字典中的资源都有一个唯一不重复的键值(Key),在标签中使用 x:Key 属性来标识它。一般来说,键值是一个字符串,但也可以用合适的扩展标签来设置为其他对象类型。非字符键值资源使用于特定的 WPF 区域,尤其是风格、组件资源及样式数据等。

10.2 资源类型

在 WPF 中的资源依赖于核心.NET 的资源系统,并在这个基础上,还添加了二进制资源和逻辑资源两种资源类型。本节中重点介绍 WPF 的二进制资源和逻辑资源。

10.2.1 二进制资源

二进制资源其实是一些传统的资源项,如位图、音频文件、视频文件和松散文件(Loose file)等。对于这些资源项可将其存储为松散文件,或者编译到程序集中。这与传统的.NET 程序其实是相同的,但在 WPF 中提供了 Resource 和 Content 两种对二进制资源的构建选项。其中,Resource 是将资源放入程序集中(如果是有本地化支持,会编译到对应语言集的子程序集中)。Content 是将这个资源作为一个松散文件加入到程序集中,程序集会记录对应的文件是否存在及其路径。这就相当于人们在 Web 开发中常用的构建动作。访问二进制资源最普通的就是对松散文件的访问,实质上和普通的.NET 应用程序是一样的。

下面介绍在 WPF 程序添加二进制资源及使用 Pack URI 路径访问二进制资源的方法。

1. 添加二进制资源

添加字符串资源(不是文件)需要使用应用程序 Properties 名称空间下的资源文件(Resources.resx),如图 10.4 所示。

Resources.resx 文件内容的组织形式是"键-值"对,编译后,Resources.resx 生成

Properties 名称空间下的 Resources 类,使用该类的方法或属性就能获得资源。XAML 编译器要访问这个类,需将默认的 Resources.resx 访问级由 Internal 改为 Public。在资源文件编译器中,添加两个名称分别是 LoginName 和 PassWord 的条目,如图 10.5 所示。接下来,在 XAML 代码和 CS 代码中访问它们。

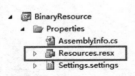

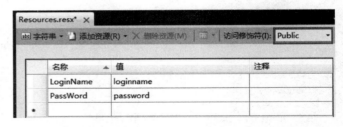

图 10.4 添加二进制资源　　　　图 10.5 设置 Resources.resx 访问级别为 Public

在 XAML 代码中使用 Resources.resx 中的资源,要将程序中的 Properties 名称空间映射为 XAML 的名称空间,再使用 x:Static 标签扩展来访问资源。XAML 代码如下。

```
<Window x:Class="BinaryResource.MainWindow"
    xmlns="http://schemas.microsoft.com/winfx/2006/xaml/presentation"
    xmlns:x="http://schemas.microsoft.com/winfx/2006/xaml"
    xmlns:prop="clr-namespace:BinaryResource.Properties"
    Title="MainWindow" Height="350" Width="525">
    <Grid>
        <Grid.RowDefinitions>
            <RowDefinition Height="20" />
            <RowDefinition Height="4" />
            <RowDefinition Height="23" />
        </Grid.RowDefinitions>
        <Grid.ColumnDefinitions>
            <ColumnDefinition Width="Auto" />
            <ColumnDefinition Width="4" />
            <ColumnDefinition Width="*" />
        </Grid.ColumnDefinitions>
        <TextBlock x:Name="loginNameTextBlock"
         Text="{x:Static prop:Resources.LoginName}"></TextBlock>
        <TextBlock x:Name="passWordTextBlock"
         Text="{x:Static prop:Resources.PassWord}" Grid.Row="2"></TextBlock>
        <TextBox BorderBrush="Black" Grid.Column="2"></TextBox>
        <TextBox BorderBrush="Black" Grid.Row="2" Grid.Column="2"></TextBox>
    </Grid>
</Window>
```

运行上述代码,页面显示效果如图 10.6 所示,在页面上显示的是图 10.5 中名称所对应的值。使用 Resources.resx 最大的优势就是便于程序的国际化、本地化。如果要将图 10.6 的运行效果页面改成中文版,只需要将图 10.5 中的资源值 LoginName 和 PassWord 改成"登录"和"密码",代码不做任何修改,再次运行代码,页面显示效果如图 10.7 所示。

如果添加的资源不是字符串,而是图片、音频或视频等,则使用外部文件作为资源,直接加入到项目文件中的资源文件夹即可。

图 10.6　运行效果

图 10.7　使用 Resources.resx 的优势

2. 使用 Pack URI 路径访问二进制资源

在 WPF 中，二进制资源添加到程序后，使用 Pack URI 路径访问二进制资源。下面分别从路径格式定义、在 XAML 代码和 CS 代码中使用 URI 路径的方法详解。

1）路径格式定义

完整的 URI 定义如下。

pack://application,,,[/可选程序集名称;][可选版本号;][文件夹名称/]文件名称

缩略后的写法如下。

[文件夹名称/]文件名称

2）在 XAML 代码中使用 URI 路径

完整路径法如下。

```
<Image x:Name = "ImageBackground"
    Source = "pack://application:,,,/Resources/Images/Car.png" Stretch = "Fill" />
```

相对路径法如下。

```
<Image x:Name = "ImageBackground" Source = "Resources/Images/Car.png" Stretch = "Fill" />
```

下面在 WPF 的项目中创建 Resources 文件夹，在该文件夹下再创建 Images 文件夹，并复制图像文件到 Images 文件夹下，如图 10.8 所示。使用 URI 相对路径的 XAML 代码如下。

```
<!-- 保留 Window 代码部分 -->
    <StackPanel>
        <Image x:Name = "ImageBackground" Stretch = "Fill"
            Source = "Resources/Images/Doraemon.png" />
    </StackPanel>
</Window>
```

运行上述代码，页面显示效果如图 10.9 所示。

图 10.8　使用外部文件做资源

图 10.9　使用外部文件做资源

3)在后台 CS 代码中使用 URI 路径

绝对路径法如下。

```
Uri imageUri = new Uri(@"pack://application:,,,/Resources/Images/Car.png",
            UriKind.Absolute);
this.ImageBg.Source = new BitmapImage(imageUri);
```

相对路径法如下。

```
Uri imageUri = new Uri(@"Resources/Images/Car.png ", UriKind.Relative);
this.ImageBg.Source = new BitmapImage(imageUri);
```

10.2.2 逻辑资源

逻辑资源是 WPF 特有的资源类型，它是存储在元素的 Resources 属性中的.NET 对象，通常需要共享给多个子元素。在此，声明 SolidColorBrush 和 LinearGradientBrush 两个对象，作为两个逻辑资源，通过静态引用方式来引用这两个逻辑资源。XAML 代码如下。

```xml
<!-- 保留 Window 代码部分 -->
    <Window.Resources>
        <SolidColorBrush x:Key = "buttonBrush">Pink</SolidColorBrush>
        <LinearGradientBrush x:Key = "backgroundBrush" StartPoint = "0,0" EndPoint = "1,1">
            <GradientStop Color = "Red" Offset = "0" />
            <GradientStop Color = "Yellow" Offset = "0.6" />
            <GradientStop Color = "Green" Offset = "1" />
        </LinearGradientBrush>
    </Window.Resources>
    <Window.Background>
        <StaticResource ResourceKey = "backgroundBrush" />
    </Window.Background>
    <StackPanel>
        <Button Margin = "5" Content = "Static Resource Button A" Background = "{StaticResource buttonBrush}" />
        <Button Margin = "5" Content = "Static Resource Button B" Background = "{StaticResource buttonBrush}"/>
    </StackPanel>
</Window>
```

运行上述代码，页面显示效果如图 10.10 所示。代码中 SolidColorBrush 和 LinearGradientBrush 通过关键字 "x:Key" 为资源命名，然后在 Button 与 Windows 的 Background 属性中通过资源名静态引用资源。

图 10.10 逻辑资源

10.3 资源引用方式

在 10.1 节和 10.2 节中，看到的案例的共性是：首先需要定义资源，才能引用资源。资源的引用有静态资源引用和动态资源引用两种方式。

10.3.1 静态资源引用

静态资源使用 StaticResource 关键字，只从资源集合中获取对象一次。静态资源引用方式不支持向前加载，故此，静态资源引用要求所有的资源应该先定义后引用。StaticResource 通常用在以下情形中。

（1）设计的 APP 是将所有的资源放入 Page 或 App 这个级别的 ResourceDictionary 中的，而且不需要在运行时重新加载。只保存一些松散文件、逻辑资源的声明等。

（2）不需要给 DependencyObject 或 Freezable 的对象设置属性。

（3）Resource Dictionary 将被编译进 DLL。

（4）需要给很多的 Dependency Property 赋值。

StaticResource 的查找行为步骤如下。

（1）检查此对象本身的 Resources 集合内是否有匹配值（根据 ResourceKey）。

（2）在逻辑树中向上搜寻父元素的 ResourceDictionary。

（3）检查 Root 级别，如 Page、Window、Application 等。

10.3.2 动态资源引用

动态资源是在运行时决定，当运行过程中真正需要时，才到资源目标中查找其值。因此，可以动态地修改它。由于动态资源在运行时才确定其值，因此效率比静态资源要低。需要说明的是，资源不仅可以在 XAML 代码中访问，也可以使用 CS 代码访问和控制它们。方法是使用 FindResource 查找资源、Resource.Add 增加资源和 Resource.Remove 移除资源。

静态资源和动态资源的区别在于静态资源只从资源集合中获取对象一次，然而动态资源在每次需要对象时都会重新从资源集合中查找对象。这意味着可以在同一键下放置一个全新对象，并且动态资源会应用该变化。

```
<!-- 保留 Window 代码部分 -->
    <Window.Resources>
        <LinearGradientBrush x:Key = "backgroundBrush" StartPoint = "0,0" EndPoint = "1,1">
            <GradientStop Color = "Red" Offset = "0" />
            <GradientStop Color = "Yellow" Offset = "0.6" />
            <GradientStop Color = "Green" Offset = "1" />
        </LinearGradientBrush>
        <ImageBrush x:Key = "TileBrush" TileMode = "Tile"
                ViewportUnits = "Absolute" Viewport = "0 0 32 32"
                ImageSource = "/Resources/Images/Snowman.png" Opacity = "0.9"/>
    </Window.Resources>
    <Window.Background>
        <StaticResource ResourceKey = "backgroundBrush" />
    </Window.Background>
    <Grid>
        <StackPanel Margin = "5">
            <Button Background = "{DynamicResource TileBrush}" Padding = "5"
                FontWeight = "Bold" FontSize = "14" Margin = "5" Content = "DynamicResource"/>
            <Button Padding = "5" Margin = "5" Content = "ChangeBrush" FontSize = "14"
```

```
                    FontWeight = "Bold" Click = "ChangeButton_Click"/>
            < Button Background = "{StaticResource TileBrush}" Padding = "5" Margin = "5"
                    FontWeight = "Bold" FontSize = "14" Content = "StaticResource"/>
        </StackPanel>
    </Grid>
</Window>
```

ChangeButton_Click 事件用来改变动态资源,其 CS 代码如下。

```
private void ChangeButton_Click(object sender, RoutedEventArgs e)
{
    this.Resources["TileBrush"] = new SolidColorBrush(Colors.LightPink);
}
```

运行上述代码,页面显示效果如图 10.11 所示。在图 10.11 中的第一个 Button 是动态引用资源,单击图中的 ChangeBrush 按钮后,页面显示效果如图 10.12 所示,DynamicResource 按钮背景色由图像笔刷变成"亮粉(LightPink)"色。

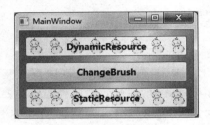

图 10.11　资源引用方式　　　　　　　图 10.12　动态资源运行中被改变

由程序执行的结果可知,静态资源引用是从控件所在的容器开始依次向上查找的,而动态资源引用是从控件开始向上查找的(即控件的资源覆盖其父容器的同名资源)。更改资源时,动态资源引用的控件样式发生变化(即"DynamicResource"发生变化)。

10.4　资　源　字　典

资源字典可以实现多个项目之间的资源共享,资源字典是一个 XAML 文档,该文档存储了所希望使用的资源。

资源字典要先创建,然后才能使用。在使用资源字典时,又分为集成资源和使用资源两个步骤。

10.4.1　创建资源字典

创建资源字典就是把需要使用的资源包含在一个 xaml 文件中。在 ResourceDictionary 中唯一的关键字"x:Key"为资源命名。

下面建立一个资源字典,将其命名为"DictionaryBackground.xaml",字典中创建 LinearGradientBrush 对象,并命名为 backgroundBrush。XAML 代码如下。

```
< ResourceDictionary xmlns = "http://schemas.microsoft.com/winfx/2006/xaml/presentation"
xmlns:x = "http://schemas.microsoft.com/winfx/2006/xaml">
```

```
    <LinearGradientBrush x:Key = "backgroundBrush" StartPoint = "0,0" EndPoint = "1,1">
        <GradientStop Color = "Red" Offset = "0" />
        <GradientStop Color = "Yellow" Offset = "0.6" />
        <GradientStop Color = "Green" Offset = "1" />
    </LinearGradientBrush>
</ResourceDictionary>
```

再创建"DictionaryString.xaml"资源字典。字典中创建 3 个字符串对象,分别命名为"str1""str2"和"str3"。XAML 代码如下。

```
<ResourceDictionary xmlns = "http://schemas.microsoft.com/winfx/2006/xaml/presentation"
    xmlns:x = "http://schemas.microsoft.com/winfx/2006/xaml"
    xmlns:sys = "clr-namespace:System;assembly = mscorlib">
    <sys:String x:Key = "str1">2016/7/10</sys:String>
    <sys:String x:Key = "str2">12:38</sys:String>
    <sys:String x:Key = "str3">28℃?</sys:String>
</ResourceDictionary>
```

10.4.2 使用资源字典

使用资源字典分为集成资源和使用资源两个步骤。集成资源是将资源字典集成到应用程序的某些资源集合中;使用资源是在集成资源之后,在当前的工程中使用这些资源。

1. 集成资源

使用资源字典,首先要将资源字典集成到应用程序的某些资源集合中。一般的做法是集成到 App.xaml 文件中。对上一节创建的两个资源字典集成到 App.xaml 中,其 XAML 代码如下。

```
<Application x:Class = "ResourceDictionary.App"
    xmlns = "http://schemas.microsoft.com/winfx/2006/xaml/presentation"
    xmlns:x = "http://schemas.microsoft.com/winfx/2006/xaml"
    xmlns:sys = "clr-namespace:System;assembly = mscorlib"
    StartupUri = "MainWindow.xaml">
    <Application.Resources>
        <ResourceDictionary>
            <ResourceDictionary.MergedDictionaries>
                <ResourceDictionary Source = "DictionaryString.xaml"/>
                <ResourceDictionary Source = "DictionaryBackground.xaml"/>
            </ResourceDictionary.MergedDictionaries>
        </ResourceDictionary>
    </Application.Resources>
</Application>
```

上述代码中的 ResourceDictionary.MergedDictionaries 属性是一个 ResourceDictionary 对象的集合,可以使用这个集合提供用户需要使用的资源。也就是说,如果需要某个资源,只需要将与该资源相关的 xaml 文件添加到这个属性中即可。

在创建资源字典的 ResourceDictionary 项目文件中,包含的文档如图 10.13 所示。

图 10.13 资源字典项目文件夹

2. 使用资源

将资源字典集成到 App.xaml 之后，就可以使用这些资源了。在 MainWindow.xaml 中设计用户的 XAML 代码如下。

```xml
<Window x:Class = "ResourceDictionary.MainWindow"
        xmlns = "http://schemas.microsoft.com/winfx/2006/xaml/presentation"
        xmlns:x = "http://schemas.microsoft.com/winfx/2006/xaml"
        xmlns:sys = "clr-namespace:System;assembly = mscorlib"
        Title = "MainWindow" Height = "350" Width = "525">
    <Window.Background>
        <StaticResource ResourceKey = "backgroundBrush" />
    </Window.Background>
    <Grid>
        <Grid.RowDefinitions>
            <RowDefinition Height = "104*"></RowDefinition>
            <RowDefinition Height = "3*"></RowDefinition>
            <RowDefinition Height = "101*" />
            <RowDefinition Height = "104*"></RowDefinition>
        </Grid.RowDefinitions>
        <Grid.ColumnDefinitions>
            <ColumnDefinition></ColumnDefinition>
            <ColumnDefinition></ColumnDefinition>
            <ColumnDefinition></ColumnDefinition>
        </Grid.ColumnDefinitions>
        <TextBlock Foreground = "White" Margin = "26,25,23,25" FontSize = "30">Date:</TextBlock>
        <TextBox Grid.Column = "1" Margin = "8,25,6,25" FontSize = "30" Name = "dateTextBox"/>
        <TextBlock Foreground = "White" Margin = "26,25,23,25" FontSize = "30" Grid.Row = "2">Time:</TextBlock>
        <TextBox Grid.Row = "2" Grid.Column = "1" Margin = "8,26,7,18" FontSize = "30" Name = "timeTextBox" />
        <TextBlock Foreground = "White" Margin = "26,25,1,25" FontSize = "30" Grid.Row = "3">Weather:</TextBlock>
        <TextBox Margin = "8,25,6,25" Grid.Row = "3" Grid.Column = "1" FontSize = "30" Name = "weatherTextBox" />
        <Button Grid.Column = "2" Margin = "20,25" x:Name = "dateButton" FontSize = "30" Click = "dateButton_Click">日期</Button>
        <Button Grid.Column = "2" Margin = "20,26,20,18" Grid.Row = "2" x:Name = "timeButton" FontSize = "30" Click = "timeButton_Click">时间</Button>
        <Button Grid.Column = "2" Margin = "20,25" Grid.Row = "3" x:Name = "weatherButton" FontSize = "30" Click = "weatherButton_Click">天气</Button>
    </Grid>
</Window>
```

上述代码中的<StaticResource ResourceKey="backgroundBrush"/>这条语句，实现对资源的引用。

下面接着在后台编写 dateButton、timeButton 和 weatherButton 3 个按钮事件的 CS 代码。

```
private void dateButton_Click(object sender, RoutedEventArgs e)
```

```
{
    dateTextBox.Text = this.FindResource("str1").ToString();
}
private void timeButton_Click(object sender, RoutedEventArgs e)
{
    timeTextBox.Text = this.FindResource("str2").ToString();
}
private void weatherButton_Click(object sender, RoutedEventArgs e)
{
    weatherTextBox.Text = this.FindResource("str3").ToString();
}
```

运行完整的代码,页面显示效果如图 10.14 所示。继续单击"日期""时间"和"天气"按钮后,页面显示效果如图 10.15 所示。

图 10.14　资源字典

图 10.15　正确使用资源字典

10.5　小　　结

本章从资源的定义出发,可以将资源定义在控件、窗体、应用程序和资源字典中。在 WPF 中资源类型有二进制资源和逻辑资源两种类型。资源的引用方式有静态资源引用和动态资源引用两种方式。资源字典要遵循先创建后使用的规则。

将资源从程序中独立出来,源程序更高效,便于管理,便于更新,维护性好,软件的自适应能力强。资源一旦定义,便可重复利用。在 WPF 中,将资源保存在 XAML 中,对设计者而言是"可见的"。

习题与实验 10

1. 练习定义复用的 SolidColorBrush 对象,再实现让登录页面的按钮和文本框的背景颜色均使用此 SolidColorBrush 对象的实例,如图 10.16 所示。熟练掌握资源可用范围的 4 个级别。

2. 利用资源字典实现如图 10.17 所示的页面。在页面上,单击"日期""时间"和"天气"3 个按钮后,页面的 3 个对应的文本框中出现相应的数值。

图 10.16 定义复用的 SolidColorBrush

图 10.17 调用资源字典页面

第 11 章

样式

前 10 章讲解了 WPF 构建应用程序的基础知识,本章开始学习 WPF 的样式。WPF 中样式的作用,就像 Web 中的 CSS 一样,为界面上的元素定制外观,以提供更好的用户界面。在 WPF 中,可以为控件定义统一的样式(Style),也可以使用样式把一组属性应用到一个或多个控件上,实现一致的主题风格。

11.1 样式的构成

样式(Styles)由设置器(Setter)、触发器(Triggers)、资源(Resources)三部分构成。

11.1.1 设置器

在设置器中创建样式的 XAML 代码如下。

```
<Style TargetType="Button">
    <Setter Property="Background" Value="Red"/>
</Style>
```

其中,TargetType 是目标类型;Property 是属性;Value 是取值。上述代码是将目标类型指向 Button 后,设置其背景色为红色。

11.1.2 样式触发器

第 9 章已重点讲解过触发器,并了解到触发器(Triggers)主要分为三类:属性触发器、数据触发器和事件触发器。在样式中,也要使用触发器,样式中使用触发器让样式的使用更加准确、灵活和高效。

样式触发器是属性或事件发生时才会被触发。下面学习如何定义和使用样式中的单属性触发器、多属性触发器、多条件属性触发器和事件触发器。

1. 单属性触发器

单属性触发器是检查从属属性的值,即元素自身属性。下面以按钮的内容属性为例,XAML 代码如下。

```
<Window.Resources>
    <Style TargetType="Button">
        <Style.Triggers>
            <Trigger Property="Content" Value="按钮">
                <Setter Property="ToolTip" Value="这是一个按钮"></Setter>
```

```
            </Trigger>
        </Style.Triggers>
    </Style>
</Window.Resources>
<Grid>
    <Button Margin = "12,12,401,270">按钮</Button>
</Grid>
</Window>
```

运行上述代码,当鼠标指针移入"按钮"时,页面显示效果如图 11.1 所示。

2. 多属性触发器

多属性触发器是检查属性的所有可能值,符合则触发。这里以按钮的内容属性为例,XAML 代码如下。

图 11.1　单属性触发器

```
<Window.Resources>
    <Style TargetType = "Button">
        <Style.Triggers>
            <Trigger Property = "Content" Value = "按钮">
                <Setter Property = "ToolTip" Value = "这是一个按钮"></Setter>
            </Trigger>
            <Trigger Property = "Content" Value = "Button">
                <Setter Property = "ToolTip" Value = "This is a button"></Setter>
            </Trigger>
        </Style.Triggers>
    </Style>
</Window.Resources>
<Grid>
    <Button Content = "按钮" Height = "23" Margin = "28,29,394,259" />
    <Button Content = "Button" Margin = "165,29,257,259" />
</Grid>
</Window>
```

运行上述代码,当鼠标指针移入"按钮"时,页面显示效果如图 11.2 所示。当鼠标指针移入 Button 按钮时,页面显示效果如图 11.3 所示。

图 11.2　多属性触发器(按钮)　　　　图 11.3　多属性触发器(Button)

3. 多条件属性触发器

多条件属性触发器是检查多个属性,属性取值都符合条件时才触发。这里以按钮的内容属性和可见属性为例,XAML 代码如下。

```
<!-- 保留 Window 代码部分 -->
<Window.Resources>
    <Style TargetType = "Button">
        <Style.Triggers>
```

```
            <MultiTrigger>
                <!-- 条件列表 -->
                <MultiTrigger.Conditions>
                    <Condition Property = "Content" Value = "按钮"></Condition>
                    <Condition Property = "Visibility" Value = "Visible"></Condition>
                </MultiTrigger.Conditions>
                <!-- 样式 -->
                <Setter Property = "ToolTip" Value = "这是一个可见按钮"></Setter>
            </MultiTrigger>
        </Style.Triggers>
    </Style>
</Window.Resources>
<Grid>
    <Button Content = "按钮" Height = "23" Margin = "28,29,394,259" />
    <Button Content = "Button" Margin = "156,29,266,259" />
</Grid>
</Window>
```

运行上述代码,当鼠标指针移入"按钮"时,页面显示效果如图 11.4 所示。当鼠标指针移入 Button 按钮时,页面无变化。

4. 事件触发器

事件触发器用来监听事件。当一个事件发生时,事件触发器就会引发相关的动画事件响应。这里以按钮的"MouseEnter"事件为例,XAML 代码如下。

```
<Window.Resources>
    <Style TargetType = "Button">
        <Style.Triggers>
            <!-- 事件触发器 -->
            <EventTrigger RoutedEvent = "MouseEnter">
                <BeginStoryboard>
                    <Storyboard>
                        <DoubleAnimation Storyboard.TargetProperty = "Opacity" From =
                            "1" To = "0.5" Duration = "0:0:3"/>
                    </Storyboard>
                </BeginStoryboard>
            </EventTrigger>
        </Style.Triggers>
    </Style>
</Window.Resources>
<Grid>
    <Button Content = "Button" Margin = "156,29,266,259" />
</Grid>
</Window>
```

运行上述代码,当鼠标指针经过 Button 按钮时,按钮的透明度在 3 秒内从 1 降到 0.5。页面显示效果如图 11.5 所示。

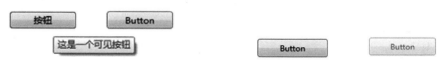

图 11.4 多条件属性触发器　　　　图 11.5 事件触发器

11.1.3 样式容器

WPF 样式可以将多个属性定义在一个样式中,而样式又存放在资源中,资源可以看作存放样式和对象的容器。

在前两节讲的样式的设置器和触发器都放在 Window 资源的< Window.Resources >与</Window.Resources >两标签之间。由此也可以得出资源是存放样式的容器。在本节不再举例说明样式容器。

在 WPF 应用程序中,通过控件的属性,可以实现更改控件的外观。但是,这种方式局限性大、不灵活且不利于维护。接下来,定义两个外观相同的 Button,XAML 代码如下。

```
<Grid>
    <Button Height="30" FontSize="18" FontWeight="Heavy" Margin="6,6,398,275">Button1
    </Button>
    <Button Height="30" FontSize="18" FontWeight="Heavy" Margin="6,42,398,239">Button2
    </Button>
</Grid>
```

上面只有两个按钮,倘若数十个按钮或者整个应用程序中所有的按钮,都这样给它们编写相同的属性,无疑很烦琐且不易维护。将上述 Button 的属性归纳起来,编写到一段样式中,为按钮指定该样式(甚至用元素类型样式时,都不需要指定按钮样式),所有按钮就具有统一样式和外观了。如果想修改按钮外观,只需要修改一下样式代码即可,所有按钮外观都会随之变化。

11.2 使用样式的方法

使用样式有多种方法,如内联样式(定义在元素内部)、已命名样式(为样式命名,使用时通过名称引用)、元素类型样式(为一种类型的元素,指定一种样式)等。

11.2.1 内联样式

内联样式是在元素定义时,在元素内部通过拓展属性 Style 来定义样式。这里定义一个 Button 控件的 Style,XAML 代码如下。

```
<Button>
    <Button.Style>
        <Style>
            <Setter Property="Button.FontSize" Value="18"></Setter>
            <Setter Property="Button.Width" Value="103"></Setter>
            <Setter Property="Button.Height" Value="30"></Setter>
            <Setter Property="Button.Content" Value="Button"></Setter>
        </Style>
    </Button.Style>
</Button>
```

内联样式的缺点在于,如果想为多个元素指定同一种样式,不得不在每个元素内部都编写同一种样式,这样不但耗费人力,并且后期维护性差。就像网页中的内联 CSS 一样。所以，一般将样式放到资源中，在元素定义时，为其指定一个样式（参见"已命名样式"）；或者在资源中，为某一类型的元素指定一个样式（参见"元素类型样式"）。

11.2.2 已命名样式

已命名样式是将相同的内联样式归纳起来，放入资源中，构成一个样式，并为其命名。在使用样式时，通过名称为元素指定该样式。

```
<!-- 保留 Window 代码部分 -->
<Window.Resources>
    <Style x:Key="Style1">
        <Setter Property="Button.FontSize" Value="18"></Setter>
        <Setter Property="Button.Width" Value="103"></Setter>
        <Setter Property="Button.Height" Value="30"></Setter>
    </Style>
</Window.Resources>
<Grid>
    <Button Style="{StaticResource Style1}" Margin="23,32,95,139">Button1</Button>
</Grid>
</Window>
```

代码中的 Button 调用"Style1"已命名样式来设置 Button 样式。在已命名样式中，还可指定目标类型、重用样式、重写样式、拓展样式。

1. 指定目标类型（TargetType）

在上述代码中，加上指定目标类型，对上节已命名样式的代码进行更改，XAML 代码如下。

```
<!-- 保留 Window 代码部分 -->
<Window.Resources>
    <Style x:Key="Style1" TargetType="{x:Type Button}">
        <Setter Property="FontSize" Value="18"></Setter>
        <Setter Property="Width" Value="103"></Setter>
        <Setter Property="Height" Value="30"></Setter>
    </Style>
</Window.Resources>
<Grid>
    <Button Style="{StaticResource Style1}" Margin="23,32,95,139">Button1</Button>
</Grid>
</Window>
```

上述代码中的 TargetType 可以修改成 Control，因为 Button 是从 Control 派生而来的。同时，CheckBox 也是派生自 Control，所以将 Style1 指定给 CheckBox 也是合适的。这样就能使多种元素共用一种样式，XAML 代码如下。

```
<!-- 保留 Window 代码部分 -->
<Window.Resources>
    <Style x:Key="Style1" TargetType="{x:Type Control}">
        <Setter Property="FontSize" Value="18"></Setter>
```

```xml
            < Setter Property = "Width" Value = "103"></Setter >
            < Setter Property = "Height" Value = "30"></Setter >
        </Style >
    </Window.Resources >
    < StackPanel >
        < Button Style = "{StaticResource Style1}" Margin = "5,20,20,5" Content = "Button1"/>
        < CheckBox Style = "{StaticResource Style1}"Margin = "0,20,20,5" Content = "Checkbox1"/>
    </StackPanel >
</Window >
```

运行上述代码，页面显示效果如图 11.6 所示，Button 和 CheckBox 共用同一样式 Style1。但是，如果在 StackPanel 布局中，再加入 TextBlock 时会出错。因为 TextBlock 不是派生自 Control，它直接继承于 FrameworkElement。加入 TextBlock，需要使用重用样式。

2. 重用样式（Reusing Styles）

重用样式是指样式可以拥有目标所没有的属性。如果定义样式，其中含有不被所有元素共享的属性，而只希望这些非共享属性应用到特定的元素上。此时，可以去掉目标类型，在属性名称前加类名。在图 11.6 中加入 TextBlock 控件，让 Button、CheckBox 和 TextBlock 3 控件的前景色为红色，字号 18，修改上述代码，使用重用样式，XAML 代码如下。

```xml
<!-- 保留 Window 代码部分 -->
< Window.Resources >
        < Style x:Key = "Style1">
            < Setter Property = "Button.Width" Value = "103"></Setter >
            < Setter Property = "Button.Foreground" Value = "red"></Setter >
            < Setter Property = "CheckBox.FontSize" Value = "18"></Setter >
            < Setter Property = "CheckBox.IsChecked" Value = "True"></Setter >
            < Setter Property = "TextBlock.Background" Value = "lightgreen"></Setter >
        </Style >
</Window.Resources >
    < StackPanel >
        <Button Style = "{StaticResource Style1}" Margin = "0,10,20,5" Content = "Button1"/>
        < CheckBox Style = "{StaticResource Style1}" Margin = "0,5,20,5" Content = "Checkbox1"/>
        < TextBlock Style = "{StaticResource Style1}" Margin = "0,0,10,5" Text = "Textblock1"/>
    </StackPanel >
</Window >
```

运行上述代码，页面显示效果如图 11.7 所示，Button、CheckBox 和 TextBlock 重用样式。将 Style1 同时指定给 Button、CheckBox1 和 TextBlock1，TextBlocak 会自动忽略不适用自身的样式属性 IsChecked。而三者公有的属性（如 Foreground、FontSize）对三者都有效。

3. 重写样式（Overriding Style）

重写样式属性类似于面向对象中的重写，其效果也类似于 CSS 中的样式覆盖。最终的外观取决于最近的样式或属性。

图 11.6　Button 和 CheckBox 共用同一样式　　图 11.7　Button、CheckBox 和 TextBlock 重用样式

现在再给一个元素指定了一个样式，其中包含 FontSize 属性值为 18。而在元素定义时，重新给它的属性 FontSize 设置了一个值 30。最终元素文本的 FontSize 将为 30。

```
<!-- 保留 Window 代码部分 -->
<Window.Resources>
    <Style x:Key="Style1">
        <Setter Property="Button.FontSize" Value="18"></Setter>
    </Style>
</Window.Resources>
    <Grid>
        <Button Width="110" Height="50" FontSize="30">Button</Button>
    </Grid>
</Window>
```

4. 拓展样式（Extending Styles）

该样式是对现有样式进行拓展，类似于面向对象中的继承或派生，可以添加新的属性或重载已存在的属性。

```
<!-- 保留 Window 代码部分 -->
<Window.Resources>
    <Style x:Key="Style1" TargetType="Button">
        <Setter Property="FontSize" Value="18"></Setter>
        <Setter Property="Foreground" Value="Green"></Setter>
    </Style>
    <Style x:Key="Style2" BasedOn="{StaticResource Style1}" TargetType="Button">
        <!-- 添加新属性 -->
        <Setter Property="FontWeight" Value="Bold"></Setter>
        <!-- 重载 -->
        <Setter Property="Foreground" Value="Red"></Setter>
    </Style>
</Window.Resources>
    <Grid>
        <Button Style="{StaticResource Style1}" Width="80" Height="30" FontSize="18"
            Margin="109,55,109,116">Button1</Button>
        <Button Style="{StaticResource Style2}" Width="80" Height="30" FontSize="18"
            Margin="109,120,109,51">Button2</Button>
    </Grid>
</Window>
```

11.2.3　元素类型样式

在页面设计时，要求用户界面上的相同控件风格要一致。以 Button 为例，页面上所有的 Button 大小相同、颜色统一等。这时定义一种元素的样式，针对一个范围内的所有元素都有效，这就是元素类型样式。

元素类型样式约束同一类型元素共享外观。如果希望一个顶级窗口内所有的元素，具有相同的样式和外观。实现方法分为以下两个步骤。

（1）在顶级窗口资源中定义一个样式，不标记 x:Key，将 TargetType 设置为一种元素类型。

（2）定义元素，不用指定 Style，窗口中所有该类型的元素，都将使用资源中定义的样式，并具有统一外观。

下面在 Window 中分别定义 Button 和 TextBlock 的样式，随后再定义 Button 和 TextBlock，其 XAML 代码如下。

```
<!-- 保留 Window 代码部分 -->
    <Window.Resources>
        <!-- Button 样式 -->
        <Style TargetType = "{x:Type Button}">
            <Setter Property = "FontWeight" Value = "Normal"></Setter>
            <Setter Property = "Foreground" Value = "Green"></Setter>
        </Style>
        <!-- TextBlock 样式 -->
        <Style TargetType = "TextBlock">
            <Setter Property = "FontSize" Value = "16"></Setter>
            <Setter Property = "Foreground" Value = "Red"></Setter>
        </Style>
    </Window.Resources>
    <StackPanel>
        <Button Name = "Button1" Width = "128" Height = "30"
            Margin = "0,10,20,5"> Button1 </Button>
        <Button Name = "Button2" Width = "128" Height = "30" Margin = "0,5,20,5">Button2
            </Button>
        <TextBlock Name = "TextBlock1" Height = "30" Margin = "50,0,10,5">TextBlock1
            </TextBlock>
        <TextBlock Name = "TextBlock2" Margin = "50,0,10,5" > TextBlock2 </TextBlock>
    </StackPanel>
</Window>
```

运行上述代码，页面显示效果如图 11.8 所示。上述代码中的元素类型样式定义在了 Window 中。用户定义的样式所在的范围决定样式所在的范围，样式除了定义在 Window 中，还可以定义在面板资源和应用程序资源中。

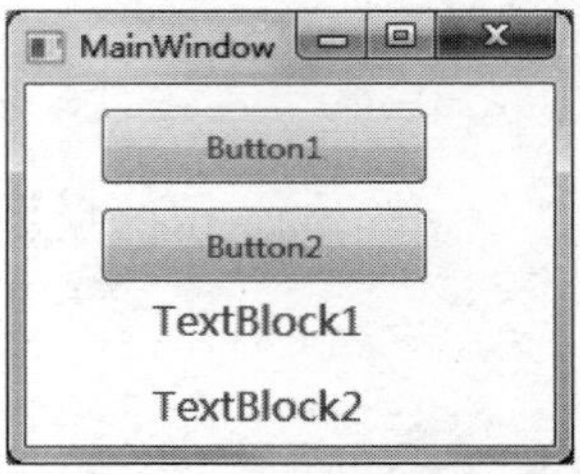

在 Grid 面板中定义 Button 样式，XAML 代码如下。

图 11.8 元素类型样式

```
<!-- 保留 Window 代码部分 -->
<Grid>
    <Grid.Resources>
        <!-- Button 样式 -->
        <Style TargetType = "{x:Type Button}">
            <Setter Property = "Foreground" Value = "Green"></Setter>
```

```xml
        </Style>
    </Grid.Resources>
    <Button Margin = "145,113,113,118">Button</Button>
</Grid>
</Window>
```

在应用程序范围,定义 Button 样式,XAML 代码如下。

```xml
<Application.Resources>
    <!-- Button 样式 -->
    <Style TargetType = "{x:Type Button}">
        <Setter Property = "Foreground" Value = "Green"></Setter>
    </Style>
</Application.Resources>
</Application>
```

11.2.4 编程控制样式

编程控制样式是通过代码更改样式。下面在 Window 中定义两个指向 Button 的样式,并定义两个 Button 元素,XAML 代码如下。

```xml
<!-- 保留 Window 代码部分 -->
<Window.Resources>
    <!-- Style1 -->
    <Style x:Key = "Style1" TargetType = "{x:Type Button}">
        <Setter Property = "FontWeight" Value = "Normal"></Setter>
        <Setter Property = "Foreground" Value = "Green"></Setter>
        <Setter Property = "Height" Value = "40"></Setter>
        <Setter Property = "Width" Value = "150"></Setter>
    </Style>
    <!-- Style2 -->
    <Style x:Key = "Style2" TargetType = "{x:Type Button}">
        <Setter Property = "FontWeight" Value = "Bold"></Setter>
        <Setter Property = "Foreground" Value = "Red"></Setter>
        <Setter Property = "Width" Value = "150"></Setter>
        <Setter Property = "Height" Value = "40"></Setter>
    </Style>
</Window.Resources>
    <Grid Width = "513" Height = "316">
        <Button Name = "Button1" Style = "{StaticResource Style1}" Margin = "30,5,323,227">Button</Button>
        <Button Name = " ChangeStyleButton" Style = " { StaticResource Style1 }" Click = "ChangeStyleButton_Click" Margin = "30,66,323,170" Height = "42">Change Button1's Style</Button>
    </Grid>
</Window>
```

在后台添加 ChangeStyleButton 的 Click 事件,CS 代码如下。

```csharp
private void ChangeStyleButton_Click(object sender, RoutedEventArgs e)
{
```

```
            this.Button1.Style = (Style)FindResource("Style2");
}
```

运行完整的代码，单击 ChangeStyleButton 按钮后，页面显示效果如图 11.9 所示，Button1 由 Style1 变成了 Style2，原来的绿色变成了红色。

图 11.9　编程控制样式

11.3　模　　板

在 11.1 节和 11.2 节中了解了样式的构成和使用样式的方法。如要自定义用户控件，通过 Style 只能改变控件的已有属性值（如颜色、字体等）。使用控件模板可以改变控件的内部结构（VisualTree 可视化树）来完成更为复杂的定制。

WPF 模板有 ControlTemplate、DataTemplate 和 ItemsPanelTemplate 3 种，它们都继承自 FrameworkTemplate 抽象类。在这个抽象类中有一个 FrameworkElementFactory 类型的 VisualTree 变量，通过该变量可以设置或者获取模板的根结点，包含外观元素树。用户自定义控件时，可以使用 ControlTemplate 来定制，这是改变 Control 的呈现，也可以通过 DataTemplate 来改变 Data 的呈现；对于 ItemsControl，还可以通过 ItemsPanelTemplate 来改变 Items 容器的呈现。其中，ControlTemplate 和 ItemsPanelTemplate 是控件模板，DataTemplate 是数据模板。

11.3.1　定制模板

WPF 的每一个控件都有一个默认的模板，该模板描述了控件的外观以及外观对外界刺激所做出的反应。用户可以自定义一个模板来替换控件的默认模板，以便打造个性化的控件。

下面定制椭圆形按钮。要求按钮的外边框颜色从白色到绿色渐变；按钮内部颜色从银色到白色渐变；当鼠标指针滑过按钮后，按钮中的文字变成红色。

分析这个椭圆形按钮的设计需求，需要替换 Button 控件默认的模板。首先，确定要使用 ControlTemplate 这个控件模板；其次，代码中要声明一个 ControlTemplate 对象，并对该对象做相应的配置；最后，再将这个 ControlTemplate 对象赋值给控件的 Template 属性即可。现在先认识 ControlTemplate 的 VisualTree 和 Triggers 两个重要属性。

（1）VisualTree，模板的可视化树，控件的外观就是使用这个属性来描述的。

（2）Triggers，触发器列表，里面包含一些触发器 Trigger，定制这个触发器列表来使控件对外界的操作做出响应，如鼠标指针经过按钮时，文字变成红色。

通过上面的分析，设计椭圆形按钮，从可视化树和触发器两方面展开设计，XAML 代码如下。

```xml
<!-- 保留 Window 代码部分 -->
<Window.Resources>
    <Style TargetType = "{x:Type Button}" x:Key = "{x:Type Button}">
        <Setter Property = "Template">
            <Setter.Value>
                <ControlTemplate TargetType = "{x:Type Button}">
                    <!-- 定义可视化树 -->
                    <Grid>
                        <Ellipse>
                            <Ellipse.Stroke>
                                <LinearGradientBrush>
                                    <GradientStop Offset = "0" Color = "White" />
                                    <GradientStop Offset = "1" Color = "Green" />
                                </LinearGradientBrush>
                            </Ellipse.Stroke>
                            <Ellipse.Fill>
                                <LinearGradientBrush>
                                    <GradientStop Offset = "0" Color = "Silver" />
                                    <GradientStop Offset = "1" Color = "White" />
                                </LinearGradientBrush>
                            </Ellipse.Fill>
                        </Ellipse>
                        <ContentPresenter Margin = "10"
                                          HorizontalAlignment = "Center"
                                          VerticalAlignment = "Center" />
                    </Grid>
                    <!-- 结束可视化树定义 -->
                    <!-- 定义触发器 -->
                    <ControlTemplate.Triggers>
                        <Trigger Property = "Button.IsMouseOver" Value = "True">
                            <Setter Property = "Button.Foreground" Value = "Red" />
                        </Trigger>
                    </ControlTemplate.Triggers>
                    <!-- 结束触发器定义 -->
                </ControlTemplate>
            </Setter.Value>
        </Setter>
    </Style>
</Window.Resources>
<Grid>
    <Button Content = "Hello, Button" Margin = "28" Name = "button1" VerticalAlignment =
        "Top" Width = "105" />
</Grid>
</Window>
```

运行上述代码，页面显示效果如图 11.10 所示。当鼠标指针滑过"Hello，Button"时，页面显示效果如图 11.11 所示。代码可分为可视化树定义和触发器定义，其中可视化树定义中包含椭圆形、椭圆形外边框的颜色、椭圆形内容的填充颜色；触发器定义实现鼠标指针经过按钮时，文字变成红色。

图 11.10　椭圆形按钮定制模板

图 11.11　鼠标滑过椭圆

11.3.2　样式与控件模板

由定制模板可知，样式与模板本质的区别是：样式只能通过改变控件的已有属性值（如颜色、字体）来简单定制控件。但是控件模板可以改变控件的内部结构（VisualTree、可视化树）来完成更为复杂的定制。

在此，设计一个按钮的数据模板，按钮控件内容是由图片和文字两部分内容组合而成的。依然将样式放在资源中，XAML 代码如下。

```xml
<!-- 保留 Window 代码部分 -->
<Window.Resources>
    <ControlTemplate TargetType="Button" x:Key="ButtonTemplate">
        <Grid>
            <Grid.ColumnDefinitions>
                <ColumnDefinition Width="100"/>
                <ColumnDefinition Width="*"/>
            </Grid.ColumnDefinitions>
            <Grid Grid.Column="0">
                <Image Source="/Images/shopping.png"></Image>
            </Grid>
            <Grid Grid.Column="1">
                <Label Content="购物车" FontFamily="微软雅黑" HorizontalAlignment=
                "Center" VerticalAlignment="Center"/>
            </Grid>
        </Grid>
    </ControlTemplate>
</Window.Resources>
    <Grid>
        <Button FontSize="40" Template="{StaticResource ButtonTemplate}"/>
    </Grid>
</Window>
```

运行上述代码，页面显示效果如图 11.12 所示。其中，"TargetType="Button""语句放在了控件模板的位置，表示使用该模板的控件类型是 Button。"x:Key="ButtonTemplate""语句命名模板资源 ID；"Template="{StaticResource ButtonTemplate}""语句是通过模板资源 ID 来调用控件模板。

当页面运行后的效果如图 11.13 所示时，则需要修改的 XAML 代码如下。

```xml
<Grid Grid.Column = "0">
    <Image Source = "/Images/page.png"></Image>
</Grid>
<Grid Grid.Column = "1">
    <Label Content = "网站首页" />
</Grid>
```

图 11.12　按钮样式模板

图 11.13　套用按钮样式模板

11.3.3　样式与数据模板

在 11.3.2 节中了解到控件模板表示控件外观的显示方式,而数据模板是数据内容的呈现方式。相同的数据内容,可以让它的表现方式多样化。数据模板适用于 Content Control 类控件与 Items Control 类控件。

在此构建 Person 类,CS 代码如下。

```csharp
namespace ButtonDataTemplate
{
    public class Person
    {
        public string Name { get; set; }
        public string Sex { get; set; }
        public string Address { get; set; }
        public string Photo { get; set; }
    }
}
```

再创建 PersonCollection,继承自 Person,作用是收集数据,并为 MVVM 模式中的 ViewModel 做准备。

```csharp
public class PersonCollection : Person
{
  System.Collections.ObjectModel.ObservableCollection < Person > persons = new System.Collections.ObjectModel.ObservableCollection<Person>();
        public PersonCollection()
        {
            persons.Add(new Person()
            {
                Address = "北京",
                Name = "北漂一族",
                Photo = "/Images/Beipiao.jpg",
                Sex = "男"
            });
```

```csharp
            persons.Add(new Person()
            {
                Address = "山西",
                Name = "高山流水",
                Photo = "/Images/Gaosanls.jpg",
                Sex = "男"
            });
            persons.Add(new Person()
            {
                Address = "河北",
                Name = "河鱼天雁",
                Photo = "/Images/Heyuhy.png",
                Sex = "男"
            });
        }
        public System.Collections.ObjectModel.ObservableCollection<Person> PersonList
        {
            get { return this.persons; }
        }
    }
```

还需要将 ViewModel 与界面建立关联关系,CS 代码如下。

```csharp
public MainWindow()
{
    InitializeComponent();
    this.DataContext = new PersonCollection();        //实现 MVVM 的核心语句
}
```

最后需要设计前台用户的 UI 界面,用数据模板和 ListBox 来显示数据,还采用数据绑定技术,XAML 代码如下。

```xml
<Window x:Class="ButtonDataTemplate.MainWindow"
        xmlns="http://schemas.microsoft.com/winfx/2006/xaml/presentation"
        xmlns:x="http://schemas.microsoft.com/winfx/2006/xaml"
        xmlns:local="clr-namespace:ButtonDataTemplate"
        Title="MainWindow" Height="350" Width="525">
    <Window.Resources>
        <DataTemplate xmlns:local="clr-namespace:ButtonDataTemplate"
                    x:Key="MyDataTemplate" DataType="{x:Type local:Person}">
            <Grid VerticalAlignment="Center" HorizontalAlignment="Center" Margin="4">
                <Grid.ColumnDefinitions>
                    <ColumnDefinition Width="Auto"/>
                    <ColumnDefinition Width="Auto"/>
                </Grid.ColumnDefinitions>
                <Image Source="{Binding Photo}" Width="150" Grid.Column="0"
                    Grid.RowSpan="1"/>
                <TextBlock Text="{Binding Name}" Grid.Column="1" Grid.ColumnSpan="1"
                        HorizontalAlignment="Center" VerticalAlignment="Center"/>
            </Grid>
```

```xml
        </DataTemplate>
    </Window.Resources>
    <Grid>
        <ListBox ItemsSource = "{Binding PersonList}"
            ItemTemplate = "{StaticResource MyDataTemplate}"/>
    </Grid>
</Window>
```

运行上述代码,页面显示效果如图 11.14 所示。观察该数据模板,并不是非常好看,下面采用样式模板重新设定,XAML 代码如下。

```xml
<Window.Resources>
    <Style TargetType = "ListBoxItem" x:Key = "ListViewItemStyle">
        <Setter Property = "Template">
            <Setter.Value>
                <ControlTemplate TargetType = "ListBoxItem" x:Name = "ListBorder">
                    <Border BorderBrush = "Red" BorderThickness = "1">
                        <Grid>
                            <Grid.ColumnDefinitions>
                                <ColumnDefinition Width = "190"/>
                                <ColumnDefinition Width = " * "/>
                            </Grid.ColumnDefinitions>
                            <Grid Grid.Column = "0">
                                <Image Source = "{Binding Photo}" Name = "Image"/>
                            </Grid>
                            <Grid Grid.Column = "1">
                                <Label Content = "{Binding Name}" x:Name = "textContent"
                                   FontFamily = "微软雅黑" VerticalAlignment = "Center"
                                       HorizontalAlignment = "Center" />
                            </Grid>
                        </Grid>
                    </Border>
                </ControlTemplate>
            </Setter.Value>
        </Setter>
    </Style>
</Window.Resources>
<Grid>
    <ListBox ItemsSource = "{Binding PersonList}"
        ItemContainerStyle = "{StaticResource ListViewItemStyle}"/>
</Grid>
</Window>
```

运行上述代码,页面显示效果如图 11.15 所示。将控件模板放入样式中,为 Grid 加了红色的边框线。

图 11.14　数据模板　　　　　　图 11.15　列表样式模板

11.3.4　列表与项目模板

11.3.3 节将控件模板放入样式中，为 Grid 加了红色的边框线。但是操作鼠标，页面没有变化。原因是在触发器中没有做效果设定。下面在 ListBox 中引入 ItemsPanelTemplate 来实现，在这里将它简称为列表模板。

针对上述需求，在触发器中修改 Border 背景色。当单击鼠标时，背景为红色；当鼠标指针滑过时，背景为天蓝色。在上节的代码基础上，省略相同的代码部分，给出有变化及增加的 XAML 代码如下。

```
<!-- 略 -->
<Border BorderBrush = "Red" BorderThickness = "1" x:Name = "SS">
<!-- 略 -->
</Border>
        <ControlTemplate.Triggers>
          <Trigger Property = "IsMouseOver" Value = "True">
            <Setter Property = "Background" Value = "SkyBlue" TargetName = "SS"/>
          </Trigger>
          <Trigger Property = "IsFocused" Value = "True">
            <Setter Property = "Background" Value = "Red" TargetName = "SS"/>
          </Trigger>
        </ControlTemplate.Triggers>
<!-- 略 -->
<Grid>
        <ListBox ItemsSource = "{Binding PersonList}"
                 ItemContainerStyle = "{StaticResource ListViewItemStyle}"/>
</Grid>
```

运行上述代码，在第一张图片上单击鼠标后，背景变成红色，页面显示效果如图 11.16 所示。当鼠标指针滑过第 2 张图片时，背景变成蓝色，页面显示效果如图 11.17 所示。

图 11.16 单击鼠标的页面效果　　　　图 11.17 鼠标指针滑过的页面效果

现在让列表项横向显示,需要重写 ListBox 的 ItemsPanel 并修改布局为 WrapPanel,省略与上述代码相同部分,给出 ListBox 控件的 XAML 代码如下。

```
<!-- 略 -->
    </Window.Resources>
    <Grid>
        <ListBox Background = "Transparent" Margin = "0,5,0,5" BorderBrush = "Transparent"
            ItemsSource = "{Binding PersonList}"
            ItemContainerStyle = "{StaticResource ListViewItemStyle}">
            <ListBox.ItemsPanel>
                <ItemsPanelTemplate>
                    <WrapPanel Orientation = "Horizontal" />
                </ItemsPanelTemplate>
            </ListBox.ItemsPanel>
        </ListBox>
    </Grid>
```

运行上述代码,页面显示效果如图 11.18 所示。因为列表控件都是从 ItemsPanel 继承而来的,ItemsPanel 模板中改用了 WrapPanel,故此列表项中的数据横向显示。

图 11.18 列表项横向显示效果

11.3.5 主题与皮肤

主题和皮肤是两个易混淆的概念。主题是指应用程序外观的数据集合。而皮肤可以随着应用程序而变化。用户通过更换皮肤(简称换肤)实现最终的用户定制。

主题通常涉及操作系统的可视化特性,它会反映在所有程序的用户界面的元素上。WPF 并没有独特的"皮肤"概念,也没有一个正式的"换肤"概念,但是它可以通过其动态资源机制、样式和模板来实现换肤功能。

把样式、模板和资源放在一起,就构成一个功能包。这个功能包定义控件完整的外观。它就是一个控件主题。主题实质上是一个资源字典。在样式或控件模板中的资源属性(Resource)都是资源字典类型(ResoourceDirectionary)。

当"皮肤"这个术语被应用到用户界面中时,其实质就是指被运用于用户界面上的所有界面元素的可视化样式。一个可"换肤"的用户界面既可以在编译时定制,也可以在运行时定制(定制皮肤)。WPF 为用户界面的"换肤"提供了强大的支持。

对于一个软件来说,在很多情形下"换肤"也许将变得非常重要。用户可以根据个人审美观念来定制自己的软件界面。当做企业级的业务时,公司开发的应用程序被分发成多种客户端,每个客户端需要拥有它自己的 Logo、颜色、字体等,此时也需要"换肤"。

下面通过 Button 实现 Grid 页面"换肤"功能。

(1) 创建一个资源字典的 XAML 文件,并定义 ResourceDictionary 对象,此处的文件命名为 SkinResourceDictionary.xaml。该 XAML 文件的代码如下。

```xml
<ResourceDictionary xmlns = "http://schemas.microsoft.com/winfx/2006/xaml/presentation"
                    xmlns:x = "http://schemas.microsoft.com/winfx/2006/xaml">
    <!-- 皮肤 1 -->
    <Style x:Key = "skin_Yellow">
        <Setter Property = "Control.Background">
            <Setter.Value>
                <LinearGradientBrush StartPoint = "0.5,0" EndPoint = "0.5,1">
                    <GradientStop Color = "Yellow" Offset = "0" />
                    <GradientStop Color = "WhiteSmoke" Offset = "0.5" />
                    <GradientStop Color = "Yellow" Offset = "1" />
                </LinearGradientBrush>
            </Setter.Value>
        </Setter>
    </Style>
    <!-- 皮肤 2 -->
    <Style x:Key = "skin_Green">
        <Setter Property = "Control.Background">
            <Setter.Value>
                <LinearGradientBrush StartPoint = "0.5,0" EndPoint = "0.5,1">
                    <GradientStop Color = "Green" Offset = "0" />
                    <GradientStop Color = "WhiteSmoke" Offset = "0.5" />
                    <GradientStop Color = "Green" Offset = "1" />
                </LinearGradientBrush>
            </Setter.Value>
        </Setter>
```

```xml
    </Style>
</ResourceDictionary>
```

(2) 在 App.xaml 文件中引入资源,App.xaml 文件如下。

```xml
<Application x:Class = "SkinDictionary.App"
        xmlns = "http://schemas.microsoft.com/winfx/2006/xaml/presentation"
        xmlns:x = "http://schemas.microsoft.com/winfx/2006/xaml"
        StartupUri = "MainWindow.xaml">
    <Application.Resources>
        <!-- 引入资源库 -->
        <ResourceDictionary>
            <ResourceDictionary.MergedDictionaries>
                <ResourceDictionary Source = "SkinResourceDictionary.xaml"></ResourceDictionary>
            </ResourceDictionary.MergedDictionaries>
        </ResourceDictionary>
    </Application.Resources>
</Application>
```

(3) 指定需要换肤的对象的 style 为 DynamicResource,XAML 代码如下。

```xml
<Window x:Class = "SkinDictionary.MainWindow"
        xmlns = "http://schemas.microsoft.com/winfx/2006/xaml/presentation"
        xmlns:x = "http://schemas.microsoft.com/winfx/2006/xaml"
        Title = "MainWindow" Height = "350" Width = "525">
    <Grid Style = "{DynamicResource skin_Yellow}" x:Name = "MyGrid">
        <Button x:Name = "ChangeSkinButton" FontSize = "20" Margin = "100" Content = "换肤"
Click = "ChangeSkinButton_Click">
        </Button>
    </Grid>
</Window>
```

(4) 编写换肤的方法,后台 CS 代码如下。

```csharp
public partial class MainWindow : Window
{
    public MainWindow()
    {
        InitializeComponent();
    }
    private int skinstyle = 1;
    private void ChangeSkinButton_Click(object sender, RoutedEventArgs e)
    {
        if (skinstyle == 1)
        {
            this.MyGrid.Style = (Style)Application.Current.Resources["skin_Yellow"];
            skinstyle = 2;
        }
        else
        {
            this.MyGrid.Style = (Style)Application.Current.Resources["skin_Green"];
            skinstyle = 1;
```

```
        }
    }
}
```

运行完整的代码，页面显示效果如图 11.19 所示。在单击图 11.19 中的"换肤"按钮后，页面显示效果如图 11.20 所示。

图 11.19　初始化黄色皮肤页面

图 11.20　换肤为绿色皮肤页面

11.4　小　　结

本章重点讲述了样式的构成、如何使用样式及模板。还涉及主题和皮肤，并演示了 WPF 实现换肤的操作步骤。

习题与实验 11

1. 编写样式 1(Style1)代码，如图 11.21 所示，当单击 Change Button1's Style 按钮后，页面显示效果如图 11.22 所示，Button1 由 Style1 变成了 Style2，原来的绿色变成了蓝色，字体变大。

图 11.21　Style1

图 11.22　Style2

2. 设计某餐厅的食品展示页面，如图 11.23 所示。页面包含食品图片和食品名称。

3. 通过 Button 实现 Grid 页面"换肤"功能：初始为白色，单击"换肤"按钮后，页面红色和蓝色切换。换肤初始页面效果如图 11.24 所示，换肤到红色背景页面效果如图 11.25 所示，换肤到蓝色背景页面效果如图 11.26 所示。

图 11.23　食品展示页面

图 11.24　换肤初始页面

图 11.25　换肤到红色背景

4. 自定义用户控件，实现如图 11.27 所示的页面效果。图 11.28 是项目的文件结构。

图 11.26　换肤到蓝色背景

图 11.27　自定义时钟控件

设计提示：

（1）Watch 类是 Control 的派生类，具有表示时间、小时、分钟、秒的 4 个 DependencyProperty，Watch 就是自定义控件。该控件也可以被其他项目引用。

（2）在当前项目中存在的 XAML 文件中使用该自定义控件方法，将此 XmlNamespace 特性添加到要使用该特性的标记文件的根元素中。添加代码示例如下。

```
xmlns:MyNamespace = "clr-namespace:WpfApplication3"
```

图 11.28　文件结构

（3）在其他项目中存在的 XAML 文件中使用该自定义控件的方法是：将此 XmlNamespace 特性添加到要使用该特性的标记文件的根元素中，添加代码示例如下。

xmlns:MyNamespace = "clr-namespace:WpfApplication3;assembly = WpfApplication3"

（4）Themes 文件夹下的 Generic.xaml 是样式定义，该样式以资源字典的形式存在。

（5）资源字典可以在当前项目中被调用，也可以被其他项目引用。

第 12 章

MVVM 设计模式

最早提出设计模式概念的是建筑设计大师亚历山大(Alexander)。1970 年,亚历山大在其著作《建筑的永恒之道》里描述了设计模式。其描述是这样的:模式是一条由 3 个部分组成的通用规则,它表示了一个特定环境、一类问题和一个解决方案之间的关系。每一个模式描述了一个不断重复发生的问题,以及该问题解决方案的核心设计。他还在另一本书《建筑模式语言》中提到了现在已经定义了 253 种模式。尽管亚力山大的著作是针对建筑领域的,实际上他的观点适用于所有的工程设计领域,其中包括软件设计领域。

12.1 软件设计模式

"软件设计模式"这个术语是在 20 世纪 90 年代由 Erich Gamma 等人从建筑设计领域引入到计算机科学中来的。

12.1.1 设计模式的概念

软件设计模式(Design Pattern)是一套被反复使用、多数人知晓的、经过分类编目的、代码设计经验的总结。使用设计模式是让代码更容易被他人理解、保证代码可靠性、程序的重用性,让软件具有高内聚、低耦合特性。

软件内聚其实是从化学中的分子的内聚演变过来的,化学中的分子间的作用力强,则表现为内聚程度高。因此说,软件中内聚程度的高低标志着软件设计的好坏。低耦合是用来度量模块与模块直接的依赖关系、感知程度,耦合的衡量标准是从低到高,一般来说,耦合度越低越好。

12.1.2 设计模式的原则

为什么要提倡设计模式呢?根本原因是为了代码复用,增加可维护性。那么怎么才能实现代码复用呢?则需要遵循设计模式的六大原则。

1. 开闭原则(Open Close Principle)

1988 年,勃兰特·梅耶(Bertrand Meyer)在他的著作《面向对象软件构造》(*Object Oriented Software Construction*)中提出了开闭原则,它的原文是这样的:"Software entities should be open for extension, but closed for modification"。它表达的意思是"软件模块应该对扩展开放,对修改关闭"。以程序需要增添新功能为例,不能去修改原有的代码,而是新增代码。这正如实现一个热插拔的效果(热插拔:灵活的去除或添加功能,不影响到原有的功能)。这样做的目的是为了使程序的扩展性好,易于维护和升级。

2. 里氏代换原则（Liskov Substitution Principle）

里氏代换原则是继承复用的基石，只有当衍生类可以替换掉基类，软件单位的功能不受到影响时，基类才能真正被复用，而衍生类可以在基类的基础上增加新的行为。以生活中的球类为例，原本是一种体育用品，它的衍生类有篮球、足球、排球、羽毛球等，如果衍生类替换了基类的原本方法，如把体育用品改成了食用品（那么软件单位的功能受到影响），就不符合里氏代换原则。其目的是对实现抽象化的具体步骤的规范。

3. 依赖倒转原则（Dependence Inversion Principle）

该原则是针对接口编程，而不是针对实现编程。以计算机系统为例，无论主板、CPU、内存和硬件，都是针对接口设计的，如果针对实现来设计，内存就要对应到针对某个品牌的主板，那么会出现更换内存时，需要把主板也换掉。这样做的目的是降低模块间的耦合。

4. 接口隔离原则（Interface Segregation Principle）

使用多个隔离的接口，比使用单个接口要好。以常用登录页面为例，注册时，编写用户模块的两个接口，比编写成一个接口要好。其目的是提高程序设计灵活性。

5. 迪米特法则（Demeter Principle）

迪米特法则也称为最少知道原则。这个原则首先是由美国 Northeastern University 的 Ian Holland 在 1987 年的秋天最早提出的，后来被 UML 的创始者之一 Booch 等普及。又因为在经典著作 The Pragmatic Programmer 详细描述，而广为人知。它是指一个实体应当尽量少地与其他实体之间发生相互作用，使得系统功能模块相对独立。当一个类公开的 public 属性或方法越多，修改时涉及面也就越大，变更引起的风险扩散也就越大。使用最少知道原则目的是降低类之间的耦合，减少对其他类的依赖。

6. 单一职责原则（Single Responsibility Principle）

该原则是由罗伯特·C.马丁（Robert C. Martin）在《敏捷软件开发：原则、模式和实践》一书中给出的。马丁表示此原则是基于汤姆·狄马克（Tom DeMarco）和 Meilir Page-Jones 的著作中的内聚性原则发展而来的。它是指一个类只负责一个功能领域中的相应职责，或者可以定义为：就一个类而言，应该只有一个引起它变化的原因。其目的是降低类的复杂性，提高可读性和可维护性。

在这六大原则中，开闭原则是面向对象设计的终极目标。其他五条原则，可以看做是开闭原则的实现方法。设计模式就是实现了这些原则，从而达到了代码复用、增加了可维护性。在衡量代码的优劣时常用"高内聚、低耦合"作为标准。此处的内聚是从功能角度来度量模块内的联系，一个好的内聚模块应当恰好做一件事，它描述的是模块内的功能联系。耦合是软件结构中各模块之间相互连接的一种度量，耦合强弱取决于模块设计模式。

12.2 MVVM 设计模式概述

随着软件的更新，在不同阶段出现了不同的设计模式，有 3 个常用的设计模式，它们按照时间出现的先后顺序，分别是 MVC、MVP 和 MVVM（Model View View Model）。本节重点介绍 WPF 中用到的 MVVM 设计模式。

12.2.1 MVVM 的由来

说到 MVVM 的由来，首先要认识一下 MVC 模式。MVC 模式把用户界面交互分拆到不同的 3 种角色中，使应用程序被分成 3 个核心部件，即 Model（模型）、View（视图）和 Controller（控制器）。MVC 通信方式如图 12.1 所示。

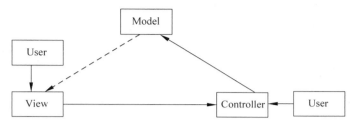

图 12.1　MVC 通信方式

由图 12.1 可知，MVC 的 Model、View 和 Controller 间通信方式是单向的，View 传送指令到 Controller；Controller 完成业务逻辑后，要求 Model 改变状态；Model 将新的数据发送到 View。接收用户指令时，MVC 可以分成两种方式。一种是通过 View 接收指令，传递给 Controller；另一种是直接通过 Controller 接收指令。MVC 3 个核心部件的功能和任务如下。

（1）Model：模型持有所有的数据、状态和程序逻辑，模型对象存取数据库中数据。模型独立于视图和控制器。

（2）View：视图用来呈现模型。视图通常直接从模型中取得它需要显示的状态与数据。对于相同的信息可以有多个不同的显示形式或视图。也就是说，视图是依据模型数据来创建。

（3）Controller：控制器位于视图和模型中间，接受用户的输入，将输入进行解析并反馈给模型，通常一个视图具有一个控制器。控制器不能处理信息，可以接受用户的请求并决定调用哪个模型部件去处理请求，然后再确定用哪个视图来显示返回的数据。

在 MVC 设计模式中，界面是主导地位，因为所有的数据全是通过界面进行控制。但是，随着时间推移，MVC 设计模式也暴露出大量缺点，因为 MVC 设计模式本质上是一个结构模式，结构模式相比行为模式而言，是静态的，相对固定的模式。随着 B/S 和互联网应用不断的普及，Web 和社会化媒体以及游戏等大量频繁交互应用普及，相对静止的 MVC 设计模式已经不适应高度交互注重行为的应用了。MVP 设计模式应运而生了。

MVP 设计模式是将 Controller（控制器）变为 Presenter（呈现器），同时将通信方式变为双向。MVP 通信方式如图 12.2 所示。

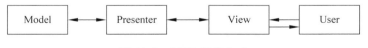

图 12.2　MVP 通信方式

在 MVP 设计模式中，Presenter 把 Model 与 View 完全分离，Presenter 与 View 是一对一的关系。Presenter 需要获取数据，并更新界面。主要的程序逻辑在 Presenter。而且是通过定义好的接口与 View 进行交互，从而使得在变更 View 时可以保持 Presenter 的不变。View 用于显示信息，具有简单的 Set/Get 方法，用户输入和设置界面显示的内容，不允许直接访问 Model。因为模型与视图完全分离，修改视图而不影响模型。将一个 Presenter 用于多个视图，而不需要改变 Presenter 的逻辑。但是，由于对视图的渲染放在了 Presenter 中，因此视图和 Presenter 的交互会过于频繁。如果 Presenter 过多地渲染视图，往往会使得它与特定的视图的联系过于紧密。一旦视图需要变更，那么 Presenter 也需要变更了。这时 MVP 的弱点就显现出来了。MVVM 设计模式的出现，弥补了 MVP 设计模式的不足。

MVVM 是将 MVP 设计模式中的 Presenter 改为 ViewModel，与 MVP 模式不一样的是，MVVM 让 View 和 View Model 双向数据绑定。这使得 ViewModel 的状态改变可以自动地传递给 View，反之亦然。在 WPF 中，通过数据绑定等技术使得数据动态更新变得简单。

12.2.2 MVVM 框架

MVVM 是微软的 WPF 带来的新的技术体验，它让软件 UI 层更加细节化、可定制化。在前面的章节中，讨论过的技术，如 Binding、Dependency Property、Routed Events、Command、DataTemplate 和 ControlTemplate，这些技术使得 MVVM 设计模式得以实现。

MVVM 框架是 MVP 与 WPF 结合的应用方式，是发展演变过来的一种新型架构。它立足于原有 MVP 框架并且把 WPF 的新特性糅合进去，以应对客户日益复杂的需求变化。MVVM 通信方式如图 12.3 所示。

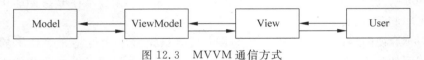

图 12.3 MVVM 通信方式

MVVM 设计模式中，用户和 View 进行交互，一个 ViewModel 可以映射多个 View。由于 View 和 ViewModel 之间有双向数据绑定关系，使数据实现动态更新。为了便于读者的理解，把 View、ViewModel 和 Model 内部结构用图 12.4 表示。

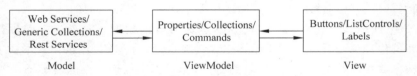

图 12.4 View、ViewModel、Model 内部结构

为了深入地理解 WPF 中 MVVM 设计模式使用的技术，对其内容结构再进一步说明如下。

（1）Model：为应用程序提供数据。其主要包含 Web Services、Rest Services、Generic Collections。

（2）ViewModel：由 Properties（属性）、Collections（集合）和 Commands（命令）3 个部分组成。这里的属性，指一个事物，它的类型可以是一个字符型，也可以是一个对象。实现接口 INotifyPropertyChanged（属性变更通知接口），那么任何 UI 元素绑定到这个属性，不管这个属性什么时候改变都能自动地和 UI 层交互。Commands 可以理解为被触发的事件，可以传递一个类型为 Object 的参数。但是前提是要实现接口 ICommand。

（3）View：主要由 3 个部分组成。第一部分，把 View Model 层的属性绑定到 Buttons（Textbox、Radio button、Togglebutton、MediaElement、Trigger an animation or ViewState change）；第二部分，把 View Model 层的集合绑定到 ListControls（ListBox、TreeView、DataGrid）；第三部分，Commands 使用 InvokeCommandAction 实现这些行为（绑定 View

Model 层的 ICommand、指出要实现的 ICommand，如 Click 事件、Selected 事件、传递参数）。

12.2.3　MVVM 的优点

MVVM 模式的主要目的是分离视图（View）和模型（Model），它有下面四大优点。

（1）低耦合。视图（View）可以独立于 Model 变化和修改，一个 ViewModel 可以绑定到不同的"View"上，当 View 变化时，Model 可以不变；当 Model 变化时，View 也可以不变。

（2）可重用性。开发人员可以把一些视图逻辑放在一个 ViewModel 中，让很多 View 重用这段视图逻辑。

（3）UI 与业务逻辑分离，实现独立开发。开发人员可以专注于业务逻辑和数据的开发（ViewModel），设计人员可以专注于页面设计（View），使用 Expression Blend 可以很容易地设计界面并生成 XAML 代码。

（4）便于测试。这里所说的测试是用代码来进行测试。对 Model、ViewModel 和 View 单元测试方便可行。

12.3　基于 MVVM 的计算器设计

实践才是硬道理。MVVM 如此之好，那怎么样的设计才算 MVVM 呢？本节将带着大家一起使用 MVVM 模式设计一款计算器。该计算器简单、轻量，却简洁明了地阐述了 MVVM 的设计思想。在这款计算器中，引入加、减、乘、除这 4 种运算功能。具有一个 Model、一个 ViewModel 和 3 个 View。并遵循先编写 Model、再设计 ViewModel、最后编写 View 的开发步骤。

首先了解计算器的整体结构，如图 12.5 所示。在图 12.5 中分别将 Command、Model、View 和 ViewModel 4 个文件夹打开后，其包含的文件结构如图 12.6 所示。

图 12.5　计算器整体结构

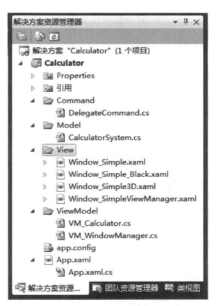

图 12.6　计算器文件结构

接下来建立 Command 文件夹下的 DelegateCommand.cs 文件。在第 9 章的学习中，了解到命令是一种逻辑约束行为，使用方法像类（封装了多种信息），可以在多个地方调用。在开发时，用户命令都继承 ICommand 接口，并实现接口的 Executed 和 CanExecute 两个方法及 CanExecuteChanged 事件。创建 DelegateCommand 类，并继承自 ICommand 接口，CS 代码如下。

```csharp
using System;
using System.Collections.Generic;
using System.Linq;
using System.Text;
using System.Windows.Input;                //新添加的引用
namespace Calculator.Command
{   //实现一个 ICommand,用于 Button 等 UI 控件的命令调用
    public class DelegateCommand : ICommand
    {
        Action execute;                  //注册方法,用于执行动作
        Func<bool> canExecute;           //或者判断是否可执行当前指令
        //构造函数
        public DelegateCommand(Action execute = null, Func<bool> canExecute = null)
        {
            this.execute = execute;
            this.canExecute = canExecute;
        }
        public bool CanExecute(object parameter)
        {
            if (canExecute == null) return true;
            return canExecute();
        }
        public event EventHandler CanExecuteChanged;
        public void UpdataCanExecuteChanged() {
            if (CanExecuteChanged!= null) CanExecuteChanged(this, EventArgs.Empty);
        }
        public void Execute(object parameter)
        {
            if (execute == null) return;
            else execute();
        }
    }
    public class DelegateCommand<T> : ICommand
    {
        Action<T> execute;
        Func<T, bool> canExecute;
        public DelegateCommand(Action<T> execute = null, Func<T, bool> canExecute = null)
        {
            this.execute = execute;
            this.canExecute = canExecute;
        }
        public bool CanExecute(object parameter)
        {
            if(canExecute == null)return true;
```

```
            return canExecute((T)parameter);
        }
        public event EventHandler CanExecuteChanged;
        public void UpdataCanExecuteChanged()
        {
            if (CanExecuteChanged!= null) CanExecuteChanged(this, EventArgs.Empty);
        }
        public void Execute(object parameter)
        {
            if (execute == null) return;
            else execute((T)parameter);
        }
    }
}
```

上述代码实现了 ICommand 接口。如果不理解,请翻阅本书 9.2 节。

12.3.1 Model

计算器的核心功能就是计算,因为此处开发的是轻量级计算器,故只保留了加、减、乘、除 4 个基本运算。在 Model 文件夹下新建业务类 CalculatorSystem(类文件 CalculatorSystem.cs),这个类只有一个方法 Calculator(计算)。该 Calculator 对文本表达式计算,并返回计算结果。本例所写出来 Calculator 这个方法即 String Calculator(String input),CalculatorSystem 类的代码如下,代码中对功能进行了注释。

```
namespace Calculator.Model
{
    public class CalculatorSystem
    { //用户自定义类型 Item,用于存放参与运算的数字及 +、-、*、/运算符
        class Item {
            public enum Type { Number, Symbol }
            public Type myType;
            public string src;
            public double Number { get { return double.Parse(src); }
                                                //文本类型转换成数值型
                                    set { src = value.ToString(); } }
                                                //数值型变成文本类型
        }
        public static String Calculator(String input) {
            List < string > dts = new List < string >();
            char[]sqs = { '+','-','*','/'};
            string temp = "";
            foreach (var v in input) {
                if (sqs.Contains(v)) {
                    dts.Add(temp);
                    dts.Add("" + v);
                    temp = "";
                }
                else {
                    temp += v;
```

```
                }
            }
            dts.Add(temp);
            List<Item> listItem = new List<Item>();
            foreach (var v in dts) {
                Item item = new Item();
                item.src = v;
                if (sqs.Contains(v.FirstOrDefault()))     //判断输入内容是数字还是运算符
                {
                    item.myType = Item.Type.Symbol;
                }
                else {
                    item.myType = Item.Type.Number;
                }
                listItem.Add(item);
            }//用于计算,运算规则,先乘除、后加减
            for (int i = 1; i < listItem.Count; i += 2) {
                if (listItem[i].src == "*") {              //乘法运算
                    listItem[i - 1].Number *= listItem[i + 1].Number;
                    listItem[i].src = "+";
                    listItem[i + 1].src = "0";
                }
                if (listItem[i].src == "/")                //除法运算
                {
                    listItem[i - 1].Number /= listItem[i + 1].Number;
                    listItem[i].src = "+";
                    listItem[i + 1].src = "0";
                }
            }
            double ret = listItem[0].Number;
            for (int i = 1; i < listItem.Count; i += 2) {
                if (listItem[i].src == "+") {              //加法运算
                    ret += listItem[i + 1].Number;
                }
                if (listItem[i].src == "-")                //减法运算
                {
                    ret -= listItem[i + 1].Number;
                }
            }
            return ret.ToString();;
        }
    }
}
```

12.3.2　ViewModel

ViewModel 文件夹下包含 VM_Calculator.cs 和 VM_WindowManager.cs 两个类文件。其中 VM_Calculator.cs 类继承 INotifyPropertyChanged(属性变更通知接口),在该类中创建一个名为 UpdateProperty() 的包装函数,其功能是实现属性变更通知。在

ViewModel 文件夹下建类 VM_Calculator.cs,代码如下。

```csharp
using System.ComponentModel;
namespace Calculator.ViewModel
{
public class VM_Calculator : INotifyPropertyChanged
{
public event PropertyChangedEventHandler PropertyChanged;
public void UpdateProperty(String propertyName)
{
        PropertyChanged.Invoke(this, new PropertyChangedEventArgs(propertyName)); }
//输出
string outputString;
public string OutputString
{
    get { return outputString; }
    set { outputString = value;}
}
//输入
string inputString;
public string InputString
{
    get { return inputString; }
    set { inputString = value; }
}
void doInput(string ch)
{
    if (ch == " = ")
    {
        outputString = CalculatorSystem.Calculator(inputString);
        UpdateProperty("InputString");            //属性改变更新界面
        UpdateProperty("OutputString");           //属性改变更新界面
        inputCommand.UpdateCanExecuteChanged();   //属性改变能否执行命令
    }
    else {
        inputString += ch;                        //添加字符串
        UpdateProperty("InputString");            //属性改变更新界面
        inputCommand.UpdateCanExecuteChanged();   //属性改变更新界面
    }
}
bool isNumber(string ch)
{
    string[] chs = { "1", "2", "3", "4", "5", "6", "7", "8", "9", "0" };
    return chs.Contains(ch);

}
bool isSymbol(string ch)
{
    string[] chs = { " + ", " - ", " * ", "/" };
    return chs.Contains(ch);
}
```

```
        bool canInput(string ch)
        {
            string lastch = null;
            if (inputString!= null) lastch = "" + inputString.LastOrDefault();
            if (isSymbol(lastch) && isSymbol(ch))return false;      //符号后面不能接符号
            if (isSymbol(lastch) && ch == " = ") return false;      //符号后面不能接符号
            return true; }
        DelegateCommand < string > inputCommand = null;
        public ICommand InputCommand
        {
            get
            {
                if ( inputCommand = = null )  inputCommand = new DelegateCommand < string >
                (doInput, canInput);
                return inputCommand;
            }
        }
        CalculatorSystem calculatorSystem;
        public CalculatorSystem CalculatorSystem {
            get { return calculatorSystem; }
            set { calculatorSystem = value; }
        }
    }
}
```

上述代码实现了 INotifyPropertyChanged。如果不理解,请翻阅本书 5.3.3 节。

在此,先来回顾 ViewModel 的作用,它连接着 View 和 Model,将用户在 View 中输入的数据传递给 Model,Model 对数据处理完成,产生的输出结果反馈给 ViewModel,ViewModel 再返回到 View 中。由此可知,ViewModel 中需要处理输入数据(string inputString)、输出数据(string outputString)、对输入数据的合法性校验方法(bool canInput (string ch),如符号后面不能接符号,View 上的按钮是否可用)、数据导入方法(void doInput(string ch),当单击 View 中的按钮后,通过此方法将数据导入 Model)、命令绑定(ICommand InputCommand,将 View 上的按钮单击事件转化为调用 doInput 方法)、属性变更通知(在 ViewModelBase 类中实现创建一个名为 UpdateProperty()的包装函数)。

接下来介绍 VM_WindowManager.cs。该类负责管理 View 中的 4 个窗口,可以实现 4 个窗体的显示和隐藏功能,该类的 CS 代码如下。

```
using Calculator.Command;
using System;
using System.Collections.Generic;
using System.Linq;
using System.Text;
using System.Windows;
using System.Windows.Input;
namespace Calculator.ViewModel
{
    class VM_WindowManager
    {   //界面
```

```csharp
public Window[] views = new Window[4];
bool[] Showings = newbool[4];
public void ShowAllView()
{
    views[0].Show(); Showings[0] = true;
    views[1].Show(); Showings[1] = true;
    views[2].Show(); Showings[2] = true;
    views[3].Show(); Showings[3] = true;
    showWindowCommand.UpdataCanExecuteChanged();
    hideWindowCommand.UpdataCanExecuteChanged();
}
bool canShowWindow(string index)
{
    int i = int.Parse(index);
    return !Showings[i];
}
bool canHideWindow(string index)
{
    int i = int.Parse(index);
    return Showings[i];
}
void doShowWindow(string index)
{
    int i = int.Parse(index);
    views[i].Show();
    Showings[i] = true;
    showWindowCommand.UpdataCanExecuteChanged();
    hideWindowCommand.UpdataCanExecuteChanged();
}
void doHideWindow(string index)
{
    int i = int.Parse(index);
    if (i == 3)
    {
        var rs = MessageBox.Show("关掉此窗口后将无法管理所有窗口是否关闭?", "警告",
        MessageBoxButton.YesNo);
            if (rs == MessageBoxResult.No) return;
    }
    views[i].Hide();
    Showings[i] = false;
    showWindowCommand.UpdataCanExecuteChanged();
    hideWindowCommand.UpdataCanExecuteChanged();
}
DelegateCommand<string> showWindowCommand = null;
public ICommand ShowWindowCommand
{
    get
    {
            if (showWindowCommand == null) showWindowCommand = new DelegateCommand
            <string>(doShowWindow, canShowWindow);
        return showWindowCommand;
```

```
            }
        }
        DelegateCommand<string> hideWindowCommand = null;
        public ICommand HideWindowCommand
        {
            get
            {
                if (hideWindowCommand == null) hideWindowCommand = new DelegateCommand
                    <string>(doHideWindow, canHideWindow);
                return hideWindowCommand;
            }
        }
    }
}
```

12.3.3 View

View 是通常意义上的 UI。虽然人们对美的认知不尽相同，但是在 UI 设计时，需要遵循 3C 原则，即 Concise（页面简洁）、Consistence（数据的一致性）、Contrast（对比度）。

鉴于页面设计中的在 3C 原则，在 View 文件夹下创建 4 个 XAML 文件，这 4 个 XAML 文件分别对应 4 个不同的视图。Window_Simple.xaml 是普通视图、Window_Simple_Black.xaml 是黑色背景视图、Window_Simple3D.xaml 是 3D 视图窗口、Window_SimpleViewManager.xaml 是管理上述 3 个视图的管理视图。

计算器 View 中的按钮形状，可以是矩形，也可以是椭圆形。分析计算器上的控件，有 Button、TextBox 和 Label 3 种类型。其中 Button 又可分为输入数据和输入运算符两种。TextBox 用于接收输入数据，Label 用于显示输出的运算结果。将所有的 Button 绑定到命令 InputCommand，但对应的命令参数不同。将 TextBox 绑定到输入字符串 InputString。将 Label 绑定到输出字符串 OutputString。在下文中对 Window_Simple.xaml 的代码给出注释，其余窗口相似，请读者注释。按照编码的顺序逐一列出这 4 个文件的 CS 代码。

（1）Window_Simple.xaml 文件的代码如下。

```xml
<Window x:Class = "Calculator.View.Window_Simple"
    xmlns = "http://schemas.microsoft.com/winfx/2006/xaml/presentation"
    xmlns:x = "http://schemas.microsoft.com/winfx/2006/xaml"
    Title = "Window_Simple" Height = "300" Width = "300" >
    <Grid>
    <!-- TextBox 输入框绑定到输入字符串 InputString -->
        <TextBox Height = "37" HorizontalAlignment = "Left" Margin = "12,12,0,0" Name =
            "textBox1" VerticalAlignment = "Top" Width = "276" Text = "{Binding InputString}" />
        <!-- Label 输出框绑定到输出字符串 OutputString -->
        <Label Height = "38" HorizontalAlignment = "Left" Margin = "195,55,0,0" Name = "label1"
            VerticalAlignment = "Top" Width = "77" Content = "{Binding OutputString}"/>
        <Grid Height = "204" HorizontalAlignment = "Left" Margin = "12,55,0,0" Name = "wrapPanel1"
            VerticalAlignment = "Top" Width = "178">
            <!-- 数字按钮"1"绑定到命令 InputCommand 命令参数为"1" -->
            <Button Content = "1" Height = "41" Name = "m1" Width = "53" Margin = "6,16,118,
                147" Command = "{Binding InputCommand}" CommandParameter = "1" />
```

```xml
        <Button Content = "2" Height = "41" Name = "m2" Width = "53" Margin = "62,16,62,
            147" Command = "{Binding InputCommand}" CommandParameter = "2" />
        <Button Content = "3" Height = "41" Name = "m3" Width = "53" Margin = "118,16,6,
            147" Command = "{Binding InputCommand}" CommandParameter = "3" />
        <Button Content = "4" Height = "41" Name = "m4" Width = "53" Margin = "6,62,118,
            101" Command = "{Binding InputCommand}" CommandParameter = "4" />
        <Button Content = "5" Height = "41" Name = "m5" Width = "53" Margin = "62,62,62,
            101" Command = "{Binding InputCommand}" CommandParameter = "5" />
        <Button Content = "6" Height = "41" Name = "m6" Width = "53" Margin = "118,62,6,
            101" Command = "{Binding InputCommand}" CommandParameter = "6" />
        <Button Content = "7" Height = "41" Name = "m7" Width = "53" Margin = "6,109,118,
            54" Command = "{Binding InputCommand}" CommandParameter = "7" />
        <Button Content = "8" Height = "41" Name = "m8" Width = "53" Margin = "62,109,62,
            54" Command = "{Binding InputCommand}" CommandParameter = "8" />
        <Button Content = "9" Height = "41" Name = "m9" Width = "53" Margin = "118,108,6,
            55" Command = "{Binding InputCommand}" CommandParameter = "9" />
        <Button Content = "0" Height = "41" Name = "m0" Width = "53" Margin = "63,155,61,8"
            Command = "{Binding InputCommand}" CommandParameter = "0" />
    </Grid>
    <WrapPanel Height = "160" HorizontalAlignment = "Left" Margin = "203,99,0,0" Name =
        "wrapPanel2" VerticalAlignment = "Top" Width = "69">
        <Button Content = " + " Height = "30" Name = "button10" Width = "68" Command = "{Binding
            InputCommand}" CommandParameter = " + " />
        <Button Content = " - " Height = "30" Name = "button11" Width = "68" Command = "{Binding
            InputCommand}" CommandParameter = " - " />
        <Button Content = "x" Height = "30" Name = "button12" Width = "68" Command = "{Binding
            InputCommand}" CommandParameter = " * " />
        <Button Content = "/" Height = "30" Name = "button13" Width = "68" Command = "{Binding
            InputCommand}" CommandParameter = "/" />
        <Button Content = " = " Height = "30" Name = "button14" Width = "68" Command = "{Binding
            InputCommand}" CommandParameter = " = " />
    </WrapPanel>
  </Grid>
</Window>
```

(2) Window_Simple_Black.xaml 文件的代码如下。

```xml
<Window x:Class = "Calculator.View.Window_Simple_Black"
    xmlns = "http://schemas.microsoft.com/winfx/2006/xaml/presentation"
    xmlns:x = "http://schemas.microsoft.com/winfx/2006/xaml"
    Title = "Window_Simple_Black" Height = "300" Width = "300" Background = "#FF101010">
    <Grid>
        <TextBox Height = "37" HorizontalAlignment = "Left" Margin = "12,12,0,0"
            Name = "textBox1" VerticalAlignment = "Top" Width = "178"
            Text = "{Binding InputString}" />
        <Label Height = "38" HorizontalAlignment = "Left" Margin = "195,12,0,0"
            Name = "label1" VerticalAlignment = "Top" Width = "77"
            Content = "{Binding OutputString}" Background = "White" />
        <Grid Height = "204" HorizontalAlignment = "Left" Margin = "12,55,0,0" Name = "wrapPanel1"
            VerticalAlignment = "Top" Width = "178">
            <Button Content = "1" Height = "41" Name = "m1" Width = "53" Margin = "6,16,118,147"
```

```xml
        Command = "{Binding InputCommand}"
            CommandParameter = "1" Background = "#FF969696" />
        <Button Content = "2" Height = "41" Name = "m2" Width = "53" Margin = "62,16,62,
            147" Command = "{Binding InputCommand}" CommandParameter = "2" Background =
            "#FF969696" />
        <Button Content = "3" Height = "41" Name = "m3" Width = "53" Margin = "118,16,6,
            147" Command = "{Binding InputCommand}" CommandParameter = "3" Background =
            "#FF969696" />
        <Button Content = "4" Height = "41" Name = "m4" Width = "53" Margin = "6,62,118,
            101" Command = "{Binding InputCommand}" CommandParameter = "4" Background =
            "#FF969696" />
        <Button Content = "5" Height = "41" Name = "m5" Width = "53" Margin = "62,62,62,
            101" Command = "{Binding InputCommand}" CommandParameter = "5" Background =
            "#FF969696" />
        <Button Content = "6" Height = "41" Name = "m6" Width = "53" Margin = "118,62,6,
            101" Command = "{Binding InputCommand}" CommandParameter = "6" Background =
            "#FF969696" />
        <Button Content = "7" Height = "41" Name = "m7" Width = "53" Margin = "6,109,118,
            54" Command = "{Binding InputCommand}" CommandParameter = "7" Background =
            "#FF969696" />
        <Button Content = "8" Height = "41" Name = "m8" Width = "53" Margin = "62,109,62,
            54" Command = "{Binding InputCommand}" CommandParameter = "8" Background =
            "#FF969696" />
        <Button Content = "9" Height = "41" Name = "m9" Width = "53" Margin = "118,108,6,
            55" Command = "{Binding InputCommand}" CommandParameter = "9" Background =
            "#FF969696" />
        <Button Content = "0" Height = "41" Name = "m0" Width = "53" Margin = "63,155,61,8"
            Command = "{Binding InputCommand}" CommandParameter = "0" Background =
            "#FF969696" />
    </Grid>
    <WrapPanel Height = "160" HorizontalAlignment = "Left" Margin = "203,99,0,0" Name =
        "wrapPanel2" VerticalAlignment = "Top" Width = "69">
        <Button Content = " + " Height = "30" Name = "button10" Width = "68" Command =
            "{Binding InputCommand}" CommandParameter = " + " Background = "#FF00FFB1" />
        <Button Content = " - " Height = "30" Name = "button11" Width = "68" Command =
            "{Binding InputCommand}" CommandParameter = " - " Background = "#FF00FFB1" />
        <Button Content = "x" Height = "30" Name = "button12" Width = "68" Command =
            "{Binding InputCommand}" CommandParameter = " * " Background = "#FF00FFB1" />
        <Button Content = "/" Height = "30" Name = "button13" Width = "68" Command =
            "{Binding InputCommand}" CommandParameter = "/" Background = "#FF00FFB1" />
        <Button Content = " = " Height = "30" Name = "button14" Width = "68" Command =
            "{Binding InputCommand}" CommandParameter = " = " Background = "#FF18FF03" />
    </WrapPanel>
  </Grid>
</Window>
```

(3) Window_Simple3D. xaml 的代码如下。

```xml
<Window x:Class = "Calculator.View.Window_Simple3D"
    xmlns = "http://schemas.microsoft.com/winfx/2006/xaml/presentation"
    xmlns:x = "http://schemas.microsoft.com/winfx/2006/xaml"
```

```xml
    Title = "Window_Simple3D" Height = "300" Width = "300">
<Grid>
    <TextBox Height = "37" HorizontalAlignment = "Left" Margin = "12,12,0,0" Name =
     "textBox1" VerticalAlignment = "Top" Width = "254" Text = "{Binding InputString}" />
    <Label Height = "38" HorizontalAlignment = "Left" Margin = "195,55,0,0" Name = "label1"
     VerticalAlignment = "Top" Width = "71" Content = "{Binding OutputString}"/>
    <Viewport3D x:Name = "view" ClipToBounds = "True" RenderOptions.EdgeMode = "Aliased"
     HorizontalAlignment = "Left" Width = "288">
        <Viewport3D.Camera>
            <PerspectiveCamera x:Name = "perspectiveCam" FieldOfView = "59" Position =
             "0.5,0,2" LookDirection = "0,0.4,-1">
            </PerspectiveCamera>
        </Viewport3D.Camera>
        <ModelVisual3D>
            <ModelVisual3D.Content>
                <AmbientLight Color = "White" />
            </ModelVisual3D.Content>
        </ModelVisual3D>
        <Viewport2DVisual3D>
            <Viewport2DVisual3D.Material>
                <DiffuseMaterial Viewport2DVisual3D.IsVisualHostMaterial = "True"
                    Brush = "White"/>
            </Viewport2DVisual3D.Material>
            <Viewport2DVisual3D.Geometry>
                <MeshGeometry3D Positions = "0,1,0   0,0,0    1,0,0    1,1,0"
                    TextureCoordinates = "0,0   0,1    1,1    1,0"
                    TriangleIndices = "0 1 2   0 2 3"/>
            </Viewport2DVisual3D.Geometry>
            <Grid>
                <Grid Height = "204" HorizontalAlignment = "Left" Margin = "12,55,0,0"
                 Name = "wrapPanel1" VerticalAlignment = "Top" Width = "178">
                    <Button Content = "1" Height = "41" Name = "m1" Width = "53" Margin =
                     "6,16,118,147" Command = "{Binding InputCommand}" CommandParameter =
                     "1" />
                    <Button Content = "2" Height = "41" Name = "m2" Width = "53" Margin =
                     "62,16,62,147" Command = "{Binding InputCommand}" CommandParameter =
                     "2" />
                    <Button Content = "3" Height = "41" Name = "m3" Width = "53" Margin =
                     "118,16,6,147" Command = "{Binding InputCommand}" CommandParameter =
                     "3" />
                    <Button Content = "4" Height = "41" Name = "m4" Width = "53" Margin =
                     "6,62,118,101" Command = "{Binding InputCommand}" CommandParameter =
                     "4" />
                    <Button Content = "5" Height = "41" Name = "m5" Width = "53" Margin =
                     "62,62,62,101" Command = "{Binding InputCommand}" CommandParameter =
                     "5" />
                    <Button Content = "6" Height = "41" Name = "m6" Width = "53" Margin =
                     "118,62,6,101" Command = "{Binding InputCommand}" CommandParameter =
                     "6" />
                    <Button Content = "7" Height = "41" Name = "m7" Width = "53" Margin =
                     "6,109,118,54" Command = "{Binding InputCommand}" CommandParameter =
```

```xml
                                            "7" />
                        <Button Content = "8" Height = "41" Name = "m8" Width = "53" Margin =
                            "62,109,62,54" Command = "{Binding InputCommand}" CommandParameter =
                            "8" />
                        <Button Content = "9" Height = "41" Name = "m9" Width = "53" Margin =
                            "118,108,6,55" Command = "{Binding InputCommand}" CommandParameter =
                            "9" />
                        <Button Content = "0" Height = "41" Name = "m0" Width = "53" Margin =
                            "63,155,61,8" Command = "{Binding InputCommand}" CommandParameter =
                            "0" />
                    </Grid>
                    <WrapPanel Height = "160" HorizontalAlignment = "Left" Margin = "203,99,
                        0,0" Name = "wrapPanel2" VerticalAlignment = "Top" Width = "69">
                        <Button Content = " + " Height = "30" Name = "button10" Width = "68"
                            Command = "{Binding InputCommand}" CommandParameter = " + " />
                        <Button Content = " - " Height = "30" Name = "button11" Width = "68"
                            Command = "{Binding InputCommand}" CommandParameter = " - " />
                        <Button Content = "x" Height = "30" Name = "button12" Width = "68"
                            Command = "{Binding InputCommand}" CommandParameter = " * " />
                        <Button Content = "/" Height = "30" Name = "button13" Width = "68"
                            Command = "{Binding InputCommand}" CommandParameter = "/" />
                        <Button Content = " = " Height = "30" Name = "button14" Width = "68"
                            Command = "{Binding InputCommand}" CommandParameter = " = " />
                    </WrapPanel>
                </Grid>
            </Viewport2DVisual3D>
        </Viewport3D>
    </Grid>
</Window>
```

（4）Window_SimpleViewManager.xaml 文件的代码如下。

```xml
<Window x:Class = "Calculator.View.Window_SimpleViewManager"
        xmlns = "http://schemas.microsoft.com/winfx/2006/xaml/presentation"
        xmlns:x = "http://schemas.microsoft.com/winfx/2006/xaml"
        Title = "Window_SimpleViewManager" Height = "305" Width = "357">
    <Grid Height = "236" Width = "312">
        <Button Content = "Hide Simple Window" Height = "23" HorizontalAlignment = "Right"
            Margin = "0,12,4,0" Name = "button1" VerticalAlignment = "Top" Width = "146" Command =
            "{Binding HideWindowCommand}" CommandParameter = "0" />
        <Button Content = "Hide Black Window" Height = "23" HorizontalAlignment = "Left" Margin
            = "162,41,0,0" Name = "button2" VerticalAlignment = "Top" Width = "146" Command =
            "{Binding HideWindowCommand}" CommandParameter = "1" />
        <Button Content = "Hide 3D Window" Height = "23" HorizontalAlignment = "Left" Margin
            = "162,70,0,0" Name = "button3" VerticalAlignment = "Top" Width = "146" Command =
            "{Binding HideWindowCommand}" CommandParameter = "2" />
        <Button Content = "Hide Manager Window" Height = "23" HorizontalAlignment = "Right"
            Margin = "0,99,4,0" Name = "button4" VerticalAlignment = "Top" Width = "146" Command =
            "{Binding HideWindowCommand}" CommandParameter = "3" Click = "button4_Click" />
        <Button Content = "Show Simple Window" Height = "23" HorizontalAlignment = "Left"
            Margin = "10,12,0,0" Name = "button5" VerticalAlignment = "Top" Width = "146"
```

```
         Command = "{Binding ShowWindowCommand}" CommandParameter = "0" />
    <Button Content = "Show Black Window" Height = "23" HorizontalAlignment = "Left" Margin =
        "10,41,0,0" Name = "button6" VerticalAlignment = "Top" Width = "146" Command =
        "{Binding ShowWindowCommand}" CommandParameter = "1" />
    <Button Content = "Show 3D Window" Height = "23" HorizontalAlignment = "Left" Margin = "10,
        0,0,143" Name = "button7" VerticalAlignment = "Bottom" Width = "146" Command =
        "{Binding ShowWindowCommand}" CommandParameter = "2" />
    <Button Content = "Show Manager Window" Height = "23" HorizontalAlignment = "Left"
        Margin = "10,99,0,0" Name = "button8" VerticalAlignment = "Top" Width = "146"
        Command = "{Binding ShowWindowCommand}" CommandParameter = "3" />
     </Grid>
</Window>
```

当运行该计算器时，4个视图窗体全部显示在页面上，如图12.7所示。在图12.7中的 Window_Simple 窗体输入"125 * 521 ="时，其余两个计算器视图呈现的数据始终与其保持一致，如图12.8所示。

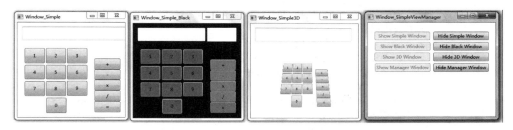

图 12.7　计算器运行初始化效果

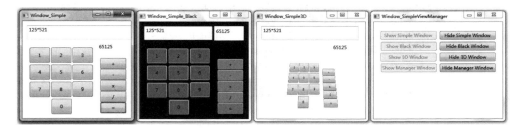

图 12.8　计算器工作页面效果

要求 View 文件夹下的4个窗体同时显示，需要设置 App.xaml 的 Startup 属性，App.xaml 文件内容如下。

```
<Application x:Class = "Calculator.App"
            xmlns = "http://schemas.microsoft.com/winfx/2006/xaml/presentation"
            xmlns:x = http://schemas.microsoft.com/winfx/2006/xaml
            Startup = "Application_Startup">
    <Application.Resources >
    </Application.Resources >
</Application>
```

App.xaml.cs 代码如下。

```csharp
namespace Calculator
{
    ///<summary>
    /// App.xaml 的交互逻辑
    ///</summary>
    public partial class App : Application
    {
        private void Application_Startup(object sender, StartupEventArgs e)
        {
            //创建计算器 View
            Window_Simple window_Simple = new Window_Simple();
            Window_Simple3D window_Simple3D = new Window_Simple3D();
            Window_Simple_Black window_Simple_Black = new Window_Simple_Black();
            //创建计算器 ViewModel
            VM_Calculator vm_calculator = new VM_Calculator();
            //创建计算器 Model
            CalculatorSystem model = new CalculatorSystem();
            //将 Model 与 ViewModel 连接
            vm_calculator.CalculatorSystem = model;
            //将 ViewModel 与 View 连接
            window_Simple.DataContext = vm_calculator;
            window_Simple3D.DataContext = vm_calculator;
            window_Simple_Black.DataContext = vm_calculator;
            //创建管理 View
            Window_SimpleViewManager manager = new Window_SimpleViewManager();
            //创建管理 ViewModel
            VM_WindowManager vm_WindowManager = new VM_WindowManager();
            //将 ViewModel 与 View 连接
            manager.DataContext = vm_WindowManager;
            //由于这个 vm_WindowManager 非常简单因此直接用 ViewModel 实现逻辑
            vm_WindowManager.views[0] = window_Simple;
            vm_WindowManager.views[1] = window_Simple3D;
            vm_WindowManager.views[2] = window_Simple_Black;
            vm_WindowManager.views[3] = manager;
            manager.Show();                                      //显示管理器界面
            vm_WindowManager.ShowAllView();                      //显示所有界面
        }
    }
}
```

细心的读者会发现，这个计算器的启动窗体不是 MainWindow。在此，解释一下 WPF 的入口。尽管 WPF 默认启动窗体是 MainWindow，但 WPF 的入口是 App.xaml 中的 App 类。WPF 程序的编译过程如下所述。

WPF 程序的编译时，首先检测是否有 App.xaml 文件，如果有 App.xaml 文件，则在 App.xaml 编译出来的 CS 文件(CS 文件的相对路径：……\obj\Debug\App.g.cs)中，添加以下入口。

```csharp
/// <summary>
/// Application Entry Point.
/// </summary>
[System.STAThreadAttribute()]
[System.Diagnostics.DebuggerNonUserCodeAttribute()]
```

```
[System.CodeDom.Compiler.GeneratedCodeAttribute("PresentationBuildTasks", "4.0.0.0")]
public static void Main() {
    WpfApplication3.App app = new WpfApplication3.App();
    app.InitializeComponent();
    app.Run();
}
```

由上述代码可知,WPF 程序的进入方式是:首先,找到 Main(),然后执行 Main()中的 app.Run()。其中,app 的属性 StartupUri 中的值即启动对象,因为前面章节所有的程序中"StartupUri="MainWindow.xaml"",所以都是把 MainWindow 作为启动该窗体。但是在本章的程序中,"StartupUri="Application_Startup"",故运行其相应的代码。

12.4 基于 MVVM 设计思想

上面采用 MVVM 设计模式实现了计算器的功能。但实际中又该如何去套用 MVVM 设计模式呢?接到一个项目时,首先划分 Model、View 和 ViewModel 结构。Model 负责数据处理业务,在本节的计算器中数据处理就是加、减、乘和除运算。ViewModel 连接着 Model 和 View,所以这里既有对输入数据的处理,又有对输出数据的显示。View 呈现用户多样的显示需求。

当然,实现 MVVM 设计模式需要用到 WPF 中的各种技术。设计 Model 业务逻辑时,用到面向对象的开发经验,因为这里是简单的数据计算,并未涉及数据库。目前大多数项目基本上都会用到数据库访问技术。ViewModel 作为数据通道,通过 INotifyPropertyChanged 确保数据的一致性。通过 CommandBinding 将 View 上的按钮单击事件转化为调用 doInput 方法。View 作为与用户的交互窗口,摆放各种 UI 控件,其中 Button、TextBox 和 Label 这 3 种控件类型使用频率最高。TextBox 输入框绑定到输入字符串 InputString;Label 输出框绑定到输出字符串 OutputString;Button 按钮绑定到命令 InputCommand。图 12.9 描述了 MVVM 设计思想。

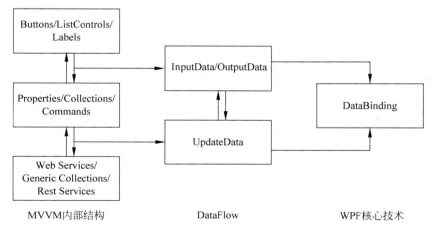

图 12.9 MVVM 设计思想

图12.9表示了MVVM设计模式、数据流及WPF核心技术这三者的关系。在项目开发中,学会利用MVVM设计模式去思考问题,使用设计模式简化项目开销,可起到事半功倍的效果。同时,还应该从软件工程管理角度来开发项目,并遵行下面的步骤。

(1) 项目的需求。分析项目中涉及的各种数据,并将这些数据分成输入数据和输出数据;同时分析系统必备的功能。

(2) 划分结构、分层设计。这里的结构正对应着MVVM设计模式。

(3) 分类管理项目文件。创建文件夹存放不同类型文件,便于后期维护。

(4) 创建用户自定义类,可以继承系统的接口,并实现其功能。

(5) 熟悉常用的接口。例如,WPF中的INotifyPropertyChanged和ICommand。

(6) 掌握开发技术。例如,WPF中的DataBinding和RoutedEvent。

12.5 小　　结

本章从软件设计模式的起源、概念和原则讲起,剖析了设计模式的发展过程。对比MVC、MVP和MVVM 3种设计模式的通信方式的差别,重点介绍了MVVM设计模式的框架、其三大组件内容结构及该模式的优点。并用基于MVVM模式的计算器实例说明WPF的技术的优势所在,阐明基于MVVM模式设计时的思考方法和项目开发步骤。

习题与实验12

1. 简述设计模式的设计概念及设计原则。
2. 对比分析MVC、MVP、MVVM 3种设计模式的通信方式的差别。
3. 了解MVVM设计模式的设计框架,简述MVVM设计模式的优点。
4. 设计一款基于MVVM设计模式的计算器,该计算器启动以后,输入"12.3 * 6"后,页面显示效果如图12.10所示。

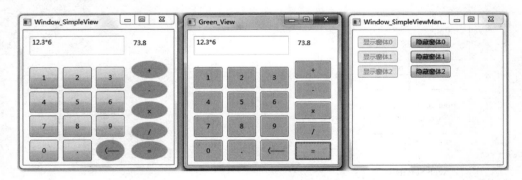

图12.10　基于MVVM的计算器

设计提示:椭圆运算符按钮用控件模板的思想实现;粉色运算符按钮用静态资源实现;本题与本章中的MVVM计算器相比,多一个小数点和退格键,注意在代码中加入业务逻辑。

参 考 文 献

[1] 陈郑军,刘振东. WPF 应用开发项目教程. 北京:中国水利水电出版社,2015.
[2] (美)Wallace B. McClure,等. C#开发 Android 应用实战——使用 Mono for Android 和.NET/C#. 王净,范园芳,田洪,译. 北京:清华大学出版社,2013.
[3] 张洪定. WPF 和 Silverlight 项目设计实例. 北京:清华大学出版社,2012.
[4] (美)Bill Evjen,Dominick Baier,Gyorgy Balassy,等..NET、C#与 Silverlight 开发圣典——分享 15 位 MVP 的最佳实践经验. 王净,范园芳,李卉,译. 北京:清华大学出版社,2012
[5] 张洪定,郭早早. WPF 和 Silverlight 教程. 天津:南开大学出版社,2012.
[6] (美)Matthew MacDonald,等. WPF 编程宝典——使用 C# 2012 和.NET 4.5(第 4 版)(.NET 开发经典名著). 王德才,译. 北京:清华大学出版社,2011.
[7] 刘铁猛. 深入浅出 WPF. 北京:中国水利水电出版社,2010.
[8] (美)Chris Anderson. WPF 核心技术. 朱永光,译. 北京:人民邮电出版社,2009.
[9] (美) Chris Sells,(英)Ian Griffiths. WPF 编程(影印版). 南京:东南大学出版社,2008.
[10] 王少葵. 深入解析 WPF 编程. 北京:电子工业出版社,2008.
[11] 程杰. 大话设计模式. 北京:清华大学出版社,2007.
[12] 刘晋钢,刘卫斌,刘晋霞. Kinect 与 Unity3D 数据整合技术在体感游戏中的应用研究. 电脑开发与应用,2014(11).
[13] Jingang Liu, Shuliang Xu. Applied Research of Weighted K-means Algorithm in Social Networks. Applied Mechanics and Materials,2014. Vol. 667.
[14] Jingang Liu. Application Research of Somatosensory Game Based on Kinect and Unity3D Data Integration Technology. Applied Mechanics and Materials,2014. Vol. 667.
[15] 微软 MSDN(网络版)WPF 部分内容. http://msdn. microsoft. com/zh-cn/library/ms754130(v=vs. 110). aspx.
[16] WPF 精品课程网址. http://moodle. cqdd. cq. cn/mod/page/view. php? id=18332.

图书资源支持

感谢您一直以来对清华版图书的支持和爱护。为了配合本书的使用,本书提供配套的资源,有需求的读者请扫描下方的"书圈"微信公众号二维码,在图书专区下载,也可以拨打电话或发送电子邮件咨询。

如果您在使用本书的过程中遇到了什么问题,或者有相关图书出版计划,也请您发邮件告诉我们,以便我们更好地为您服务。

我们的联系方式:

地　　址: 北京海淀区双清路学研大厦 A 座 707

邮　　编: 100084

电　　话: 010-62770175-4604

资源下载: http://www.tup.com.cn

电子邮件: weijj@tup.tsinghua.edu.cn

QQ: 883604(请写明您的单位和姓名)

用微信扫一扫右边的二维码,即可关注清华大学出版社公众号"书圈"。

书 圈